T0240848

Carl Adam Petri

Carl Adam Petri

Einar Smith

Carl Adam Petri

Life and Science

 Springer

Einar Smith
Fraunhofer-Gesellschaft SCAI-
Institute for Algorithms and Scientific Computing
Sankt Augustin, Germany

Translated from the German by the author and Tim Denvir.
Title of the orginal German Edition: Carl Adam Petri - Eine Biographie
© Springer-Verlag Berlin Heidelberg 2014

ISBN 978-3-662-51683-6   ISBN 978-3-662-48093-9   (eBook)
DOI 10.1007/978-3-662-48093-9

Springer Heidelberg New York Dordrecht London
© Springer-Verlag Berlin Heidelberg 2015
Softcover re-print of the Hardcover 1st edition 2015

Printed on acid-free paper

Springer-Verlag GmbH Berlin Heidelberg is part of Springer Science+Business Media
(www.springer.com)

Foreword

Whoever had the privilege of meeting Carl Adam Petri in private or public conversation remembers a very modest, humble person willing to patiently listen to his guest. When it came to his view, however, he was an adamant visionary who never allowed himself to sidetrack from his ultimate aim: to establish a comprehensive formal basis for informatics. In the cumbersome age of paper tape and punched cards, Petri rightly predicted a central usage of today's computing devices in his PhD thesis, 'Communication with Automata'. In times when only stand-alone, one-processor computers were available, Petri suggested concurrency as a fundamental phenomenon of discretely evolving systems.

Carl Adam Petri is renowned for his invention of what is known as 'Petri nets' all over the world, with places to contain tokens that are moving along transitions. Petri himself considered this concept just a starting point for a far more comprehensive theory of informatics. During his professional life, Petri carefully observed the quickly evolving world of informatics, from its inception in the 1950s until the first decade of the new millennium. He early envisaged some of the later outcomes; others he considered irrelevant. He always was missing a comprehensive discussion of the formal, theoretical basis of informatics. He never agreed with the narrative of adopting the concept of computable functions over sequences of symbols as the only fundamental basis of informatics, as frequently suggested, in particular during the hype of the 'Turing year', 2012. Instead, he had a clear vision of theoretical concepts, in accordance with the laws of physics, information flow, and information processing.

You may wish to learn more about the person with such autonomous spirit. This is what this book offers: An inspiring biography, illuminating both the personal and the professional evolution of Carl Adam Petri. Knowledgeable readers may find new perspectives to Petri nets; other readers may learn what Petri nets are about in the first place.

The author, Einar Smith, has been the most close collaborator and friend, especially of the late Carl Adam Petri. Nobody else could have better selected and described the highlights of Petri's life as a scientist.

I am very happy that Einar has compiled this text. This book adds valuable insights into an exciting aspect of the short history of informatics. I wish this book wide recognition in the informatics community and beyond.

Berlin, Germany   Wolfgang Reisig
June 2015

Preface

Carl Adam Petri passed away at the age of 83 in 2010. He spent most of his professional life as head of institute in the Gesellschaft für Mathematik und Datenverarbeitung (GMD, German National Center for Mathematics and Computer Science) in Sankt Augustin near Bonn, Germany.

In recognition of his contributions to the GMD, to basic research in computer science, and to computer science in general, the management board of the Fraunhofer Gesellschaft (in English known as the Fraunhofer Institute), with which the GMD merged in 2001, initiated a project to document Petri's life and works. Central to this project was the preparation of a biography.

I am very grateful that I was asked to undertake that task, because in this way I got the opportunity to pass on at least a small part of what I have learned from my revered teacher and friend Carl Adam. In innumerable long conversations, he explained to me, with great patience, his radical and visionary understanding of informatics and computer science and showed me the route to my own research. Also after his retirement, even when he was already heavily plagued by serious illness, he continued to share his insights with me.

The worldwide dissemination of the Petri nets, named after him, has developed a dynamic of its own; there is no end in sight, neither for theoretical research nor practical applications. This gives me the opportunity to concentrate here on the "history of ideas" behind the origins and background of nets and also on the person Carl Adam Petri himself.

This history includes research insights and approaches that have influenced net theory, even if the connections are not always obvious. If one considers Petri nets as fungi, we shall then be mainly concerned with the generating mycelium.

The main guideline we shall follow is the chronology of the principal character; however, whenever the content requires it, we give precedence to a clarification of the thoughts and their interrelations.

In compiling this biography, besides my own conversations with Carl Adam Petri, the contributions of his close collaborators Hartmann Genrich and Wolfgang Reisig were also of great use. I am also very grateful for all the helpful personal background information that Carl Adam's son Tobias gave me.

I was also able to profit extensively from written contributions from Tobias Petri, who for instance has prepared an entire CD with data, stories, and photos of his father. The text of a laudatory speech delivered by a colleague and friend of Carl

Adam's, Lu Ruqian from the Chinese Academy of Sciences, on the occasion of Petri's 60th birthday was also very useful.

The most important material was, however, a loose collection of more than 300 handwritten A4 pages that Petri perhaps intended to use himself in a future autobiography. Where there were still some details missing, I made extensive use of the Internet, to the development of whose predecessor, the ARPA-net, Petri contributed in the 1960s.

This book is intended both for readers with previous knowledge in computer science, whether including Petri nets or not, and for "interested nonprofessionals," who might perhaps develop an interest in a further study of Petri nets. Lastly, it is also intended for readers who simply want to get to know a remarkable personality of contemporary science. The text is organized in such a way that skipping over formal details should affect the overall understanding as little as possible.

The text is a slightly revised translation of the German original *Carl Adam Petri. Eine Biographie*, published by Springer-Verlag in 2014. Whenever it seemed necessary, however, I have added additional notes that could be useful to readers not familiar with Germany and the German language.

The translation was a two-phase process. First, I made a raw version myself, which was then corrected and refined by a native English speaker, Tim Denvir, a software engineer, computer scientist, and a regular consultant and book reviewer for Springer. In fact, Denvir's contributions were not limited to the linguistic level. In many cases, his critical remarks helped me to clarify the formal argument. However, I myself am responsible for any remaining idiomatic idiosyncrasies.

For further studies of Petri's life and works, the reader is referred to the Deutsche Museum in Munich, where Petri's scientific estate, donated by his son Tobias, has been meticulously curated and archived. There Petri is in the best of company, for instance with Ernst Mach and not least Konrad Zuse.

For stimulation and encouragement, I would in particular like to thank Tobias Petri and Wolfgang Reisig from the Humboldt University, Berlin. I am also grateful to Tobias Petri for his permission to use drawings and photographs from his father's private archive.

Wolfgang has supported the publication energetically, not least by establishing contact with Springer-Verlag and by offering to write a foreword.

Finally, I would like to thank all those involved at Springer-Verlag, especially Hermann Engesser and Dorothea Glaunsinger, for their friendly and constructive cooperation.

Sankt Augustin, Germany   Einar Smith
June 2015

Contents

Introduction

1

Petri nets are a world-renowned means of modeling distributed systems, with applications in vastly different sectors such as banking, economics, telecommunication systems, workflow management, conflict resolution, process-control, biochemistry and system biology.

Through their easily understood principles and intuitive graphical descriptions, Petri nets can visually depict and explain complex structures and relationships also to non-specialists. On the other hand, they offer profound mathematical analysis methods to the expert.

Less well known is the fact that the founder of net modeling, Carl Adam Petri, developed most of the central ideas of his theory during his work at the Gesellschaft für Mathematik und Datenverarbeitung (GMD, German National Center for Mathematics and Computer Science) in the small town of Sankt Augustin near Bonn, at that time capital of Western Germany. Petri's personal development, his works and achievements within the GMD until his retirement—and also beyond—are the subject of this biography.

The reader does not require any previous knowledge of Petri nets, but a basic grasp of the fundamental concepts will of course facilitate his/her understanding. In the present chapter we provide a short introduction. How short? To cite one of Petri's favorite poets, Bob Dylan (the others were Sappho, Alcaeus and Horace): "And just how far would you like to go in? Not too far but just far enough so's we can say that we've been there."

© Springer-Verlag Berlin Heidelberg 2015
E. Smith, *Carl Adam Petri*, DOI 10.1007/978-3-662-48093-9_1

1.1   Petri Nets

The basic idea behind Petri nets is quite simple: A net consists of *places* (represented by circles), which can be marked by a token to indicate the *holding* of a possible state, and *transitions* (represented by squares), which permit a change of the state. According to context, the term "condition" is also often used for "place". Referring to a condition, we may also say that it "is satisfied" instead of it "holds". In such contexts then often also the term "event" is preferred over "transition".

Example

Figure 1.1a shows a situation in which a lamp is turned off, represented by a token on the place "off". The transition "goes on" is *enabled* because the input-place is marked. Through the *occurrence* of the transition the token is withdrawn, and the output-place "on" is marked instead.

Figure 1.2 shows Petri's son Tobias as observer of the system in Fig. 1.1.

In the example above the transition "goes on" has only *one* input- and *one* output-place. The power of the nets, however, arises from the fact that there may be more than one. In general, to *enable* a transition, *all* input places must be marked. The transition then withdraws the tokens from the input-places, and marks the output-places instead.

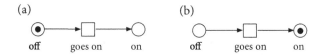

Fig. 1.1 (a) shows the situation before the occurrence of the transition "goes on", (b) the result

Fig. 1.2 Fiat lux! Drawing by C. A. Petri

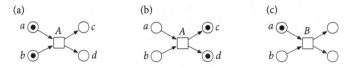

Fig. 1.3 In (**a**) A is enabled. Occurrence leads to (**b**). In (**c**) B is not enabled

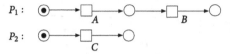

Fig. 1.4 Sequential execution of A and B in P_1, and a concurrent execution of C in P_2

Example

In Fig. 1.3a the transition A is enabled, because both preconditions a and b are satisfied. Figure 1.3b shows the result after the occurrence of A. Now both output-places c and d are marked. The transition B in Fig. 1.3c is not enabled, since only a is satisfied, but not b.

1.2   Concurrency, Sequence and Conflict

The characteristic feature of Petri's approach is to consider *concurrency* as a fundamental phenomenon. Let us take a look at a typical example:

Example

Consider a system of two processors P_1 and P_2 as shown in Fig. 1.4. Assume that P_1 executes A and B *sequentially*. This induces a structural relationship "B after A". Assume that, independently of P_1, an action C is executed by P_2. Then C is *concurrent* to both A and B. This structural independence remains, even if a possible observer notes a temporal order A-C-B.

Moreover, with increased spatial distance between the system components, the notion of temporal order loses more and more of its relevance, whereas notions such as *structural dependence* and *independence* become more significant.

Besides sequence and concurrency, there is a third possible relationship between transitions, that of a *conflict*: Two transitions are in conflict, when the occurrence of one of them disables the other.

Example

Such a situation is illustrated in Fig. 1.5. Here *either* A or B can occur, but not both. If, say, A occurs, then the necessary precondition for B no longer holds.

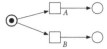

Fig. 1.5 Conflict. Occurrence of one of the transitions *A* or *B* disables the precondition for the other

Fig. 1.6 Fundamental situations. (*al* denotes the relation of *alternative*, *co* denotes *concurrency*, *li* is short for *line* to denote sequence, *sm* stands for *similarity*)

The above three relationships between transitions—sequence, concurrency and conflict—are in fact sufficient to describe all fundamental situations in Petri nets. Figure 1.6 shows a drawing by Petri, in which he illustrates these phenomena. (The *likeness*-relation is not of interest in the present chapter. It will be explained later in Chap. 10.) Most importantly, it is the concept of *concurrency* that makes up the *differentia specifica* to classical system models, in which process flow is generally considered to be ordered along a totally ordered time line. A large part of Petri's work, known as "concurrency theory", is concerned with the fundamentals and laws of concurrency, causal dependence and independence.

1.3   The Four Seasons

Figure 1.7 shows a somewhat larger net. It is often called the *Four Seasons*, because in illustrations in the literature, the states in the four corners are in fact often interpreted as the seasons of the year.

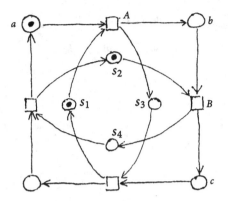

Fig. 1.7 Four Seasons. Hand-drawn sketch from [16]

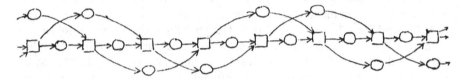

Fig. 1.8 Process of a run in the Four Seasons system

Petri also often uses this net to illustrate certain aspects of quantum-mechanical oscillators. However, it is probably best known as the logo for various international conferences and publication series.

In the situation depicted, the transition A is enabled, because the input-places a and s_1 are marked. Occurrence of A withdraws these tokens and instead marks the places b and s_3. The token on s_2 is not affected by this action. Now, in the resulting constellation, B is enabled, and an occurrence of B will transform a marking of b and s_2 into one of c and s_4, where this time the token on s_3 remains unaffected.

Figure 1.8 shows a so-called *process* of the Four Seasons system. Basically it results from moving the tokens according to the transition rule, and recording a corresponding unfolding of the net as a protocol of the system occurrences.

1.4 Distributed Access

To conclude our brief introduction, we present a somewhat larger example, in which all of the phenomena considered so far interact in a manner typical of distributed systems. Figure 1.9 shows a model to control the access to a shared resource, say a printer, by two agents a and b.

The system has to ensure that at any time *only one* user has access, and—on the other hand—that every user that requires the printer will eventually get it. Currently both agents are in a state where they do not request access, represented by the tokens

Fig. 1.9 Controlled access to a printer shared by two agents a and b

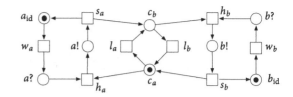

Fig. 1.10 Non-sequential process of the system in Fig. 1.9

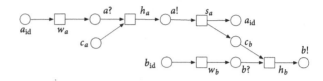

on the places a_id and b_id (The tag "id" is short for "idle"). In this situation a, say, can issue a request. In the model this is represented by the occurrence of w_a, after which the condition $a?$ is satisfied (depicted graphically by displacing the token).

The printer access control is based on a mechanism that alternately polls the demands of the agents. This is represented by the inner circle c_a-l_a-c_b-l_b. The token on c_a indicates that the printer is currently offered to a because now h_a is enabled. Occurrence of h_a withdraws the two input-tokens and puts a token on $a!$, reserving exclusive access for a. The transition s_a terminates the use; a returns to the state "idle", c_b is marked, and the printer is now ready for user b.

We shall not go deeper into the technical details of the model in this chapter; here its purpose is only to illustrate the fundamental phenomena in nets. The reader is encouraged to identify situations of concurrency, sequence and conflict. For the curious reader, we mention that we return to the more subtle problems of mutual exclusion in Sect. 6.5 and 10.9.

Figure 1.10 illustrates a possible run of the system. Formally, the system-run is represented by a so-called *non-sequential process*. Concurrency and structural dependency are directly reflected in the process. Conflicts, on the other hand, do not show up in processes, since within an actual system run, conflicts have to be *resolved*, so that for each conflict only *one* branch appears. Tokens are not required in a process, since each instance of a place in a process diagram already represents the marking of that place. In the process considered here, for example, there are *two* instances of a_id.

1.5   Literature

Worldwide there is a vast literature on Petri nets, both theoretical and applied. In the 1980s an attempt was made to establish a comprehensive bibliography. Today this would probably amount to an interminable enterprise.

Fig. 1.11 An interactive
board game

For further reading, the book *Understanding Petri Nets. Modeling Techniques, Analysis Methods, Case Studies* [23] (or the German original [22]) by Wolfgang Reisig can be recommended. Reisig is a long-standing close collaborator of Petri's.

1.6   The Petri Puzzle

In 1997 Petri was awarded the *Werner von Siemens Ring* (more about this in Sect. 11.5). On this occasion, Hartmann Genrich, another one of Petri's closest collaborators for many years, developed the Petri puzzle shown in Fig. 1.11, with which various aspects of nets could be practiced in a playful manner.

From the game instructions: "This Petri puzzle is a game for enjoyment and learning. It has the form of a board game, and involves the cooperation of an unlimited number of actors in a system."

The rules deal with the token game in nets, but also, interestingly, contain instructions on "how to extend or reduce the game board *during* a game." We shall return to the idea behind this rule in Sect. 4.1.

Unfortunately, the number of copies produced was very limited, so that today the game is kept as a rare and treasured collector's item. However, it may well be reissued as an interactive game in the Internet, in the foreseeable future.

Infancy and Youth

2

Contents

Carl Adam Peter Petri was born on July 12, 1926, in Leipzig, Germany. The name can be traced back to a Swedish ancestor, Olof Pettersson (1493–1552), who latinized his name to Olaus Petri. Olaus Petri is considered to be one of the most important pioneers of the protestant reformation in Sweden. In the sixteenth or seventeenth century a branch of the family emigrated to Germany. Many of Carl Adam's ancestors were protestant pastors in northern Germany. This line of ancestors also explains the spelling of the first name Carl with C, which in contrast to normal German spelling with K, is very common in Sweden.

Carl Adam's father Max Petri (1888–1972) was a learned man. After a classical grammar school education at the Thomasschule in Leipzig, he studied mathematics in Lausanne, Kiel and Leipzig. In 1914 he received a PhD *summa cum laude* from the University of Leipzig for a thesis on algebraic geometry. He spoke several languages, and devoted himself to the study of comparative linguistics. During the first world war he was employed by the German army as an interpreter at the Russian front.

Carl Adam's mother Elfriede, née Dietze, (1896–1970) worked as a seamstress. She descended from a family of craftsmen from Thuringia.

© Springer-Verlag Berlin Heidelberg 2015

E. Smith, *Carl Adam Petri*, DOI 10.1007/978-3-662-48093-9_2

2.1   Preschool

Carl Adam grew up as an only child. The family lived in a narrow three-room flat in one of Leipzig's suburbs, Volkmarsdorf, Ludwigstraße 76. The family's economic conditions were modest. The father periodically earned his living through sporadic jobs such as the delivery of postal parcels. After Carl Adam's grandfather, Max Sr., died in 1922, his father helped out in the family firm for some years. This small business produced bookbinding materials, for which there was a not insignificant demand in Leipzig, renowned as a 'city of books'. His grandmother, Luise Petri (née Oehlkers, 1863–1957), took over the firm after her husband died. Carl Adam's mother contributed to the household income by taking on sewing jobs.

Only after the death of his father in 1972 did Carl Adam first become aware of one additional, and probably not entirely negligible, factor in the family's meager financial circumstances: It turned out that his father had already been married before, and still had to pay alimony for his divorced wife and a daughter. According to Tobias Petri, this could in fact have been the reason, why Carl Adam's father was reluctant to aspire to well paid work, since he would anyway have had to hand the earnings over to his former family. Later, opportunities for a profitable career had passed by.

At the end of the 1930s Max Petri became an actuary for an insurance company. During the second world war he was employed by the Rheinmetall company, known as supplier of automotive parts and military technology, among other things. There he worked as consultant for the procurement of tabulating machines, and performed numerical computations for the armed forces. After the war he became a grammar-school teacher in mathematics, ancient languages, and Russian, which was the most important subject, since Leipzig by then had become part of the Russian occupied zone.

Intellectual Influences The intellectual development of young Carl Adam, however, was not impaired by the family's difficult economic circumstances. Leipzig had for centuries been a city of science and culture, and remained comparatively liberal and open-minded up to the beginning of the second world war. A large circle of his father's colleagues and friends, consisting of merchants and academics from all over the world, could still meet undisturbed at the Petris'. The home library had

Fig. 2.1 Occasionally Carl Adam had to assist his mother in her tailoring work. Drawing by Petri around 2002

benefited from an in truth rather regrettable fact: Due to the world economic crisis, many bookbinders had gotten into financial difficulties, and now paid for binding-materials with books rather than money. Thus, already as a child, Carl Adam was confronted with manifold ideas, languages and books. He had learned to read at the age of three, and in his father's intellectual environment, he developed a particular interest in physics and chemistry.

From his mother's side, on the other hand, the intellectual influences were rather limited. She was not well educated and her background was a mixture of naive Christian belief and popular superstition. In this force field between rational and mystical thinking, Carl Adam searched for explanations, for instance of the miracles in the Old Testament. This endeavor got to the point, that his mother began to hope that her son would later decide to become a theologian.

However, Carl Adam chose the path of rationality. As he later recalls: "When I was four, I noticed that my mother used an Egyptian dream book to regulate the family life. When regulation became oppression, I threw the book into the stove and watched the flames consume it. I was hated for this, and was told what sin is. But as my father smiled soon after, I recovered."

This episode however did not remain without consequences: "So it came that I read, at the age of five, Freud's popular works on dreams, on 'Fehlleistungen', and on jokes—of course in secret. I was deeply moved, and became afraid of the adults, because they appeared to be able to look right through me."

This discomfort towards psychology is without doubt rather widespread, but in the case of Carl Adam, it had a remarkable consequence. As he recalls: "At six I was tested for admission to school. I did not pass the first test, and was rejected for lack

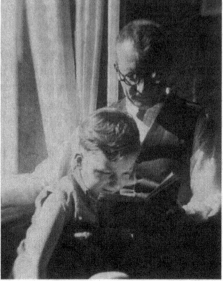

Fig. 2.2 The young Carl Adam, *on the right*, with his father

Fig. 2.3 Carl Adam with Yes-sticker. Drawing by Petri, around 2002

of intelligence. I had been shown a square piece of lacquered paper, and the director had asked what I could say about it. Sensing a psychological trap, and not believing that a straight answer would satisfy him, I had judged the situation pragmatically, that my low status required politeness in the first place; so my answer had been 'schön!'—beautiful. Thus I gained a full year of my life, for playing and reading."

2.2   Machtergreifung: The Nazi's Seizure of Power

On January 30, 1933, Adolf Hitler was proclaimed Chancellor of the Reich. On March 23, the Reichstag passed the so-called "Ermächtigungsgesetz" (Enabling Act), which established the national-socialist dictatorship. In a national referendum Hitler decided to let the people's voice "confirm" his absolute power. "The Führer calls upon you! So fulfill your duty" was proclaimed from the election posters. To avoid reprisals the Petris went to vote. To avoid assaults by the Nazi-mob they brought little Carl Adam along, whom they had equipped with a large "Yes"-sticker.

They could approve in public but rebel in private, by secretly making fun of the rulers. At home the family had a talking parrot and a canary, whom they had given the names Dölfi and Josef, alluding to Adolf Hitler and Josef Goebbels. Fortunately this was never overheard by any malicious people.

2.3   Elementary School

In 1933 Carl Adam was finally admitted to school at the age of seven. He proved to be highly gifted, so that the school's director apologized for his initial misjudgment, and let Carl Adam skip a grade. Anyone who knew Petri can confirm that the director was not really to blame. In later life too, Petri's remarks were often of a kind that left the listener in doubt whether it was a trivial observation or a statement of profound insight.

2.4   Grammar School

After elementary came grammar school. As his father before him, Carl Adam was admitted to the venerable Thomasschule, a school famous for its education, specializing in classical humanities and music. Indeed, centuries ago Johann Sebastian Bach taught at this Thomasschule. In the entrance examinations, Carl Adam benefited from the fact that he had already acquired some knowledge of Latin. But the examination also included a musical element. It thus proved advantageous that Carl Adam already played the violin, piano and flute. Besides a knowledge of instruments and a good voice, the candidate was required to sing a short sequence of tones, which was only played to him once on the piano. Carl Adam also rose to this challenge.

There remained the question of the school fee, which the family would not be able to provide. However, Carl Adam was among the three best students of his year, and received a scholarship from the city of Leipzig in January 1936.

2.5   First Encounter with the Theory of Knowledge

As a reward his father gave him a 400 page book *Ergebnisse und Probleme der Naturwissenschaften* (Results and Problems in Natural Sciences) by the physicist and natural philosopher Bernhard Bavink, probably a preprint from the bankruptcy of the printer. Carl Adam devoured the book (in particular since—as he emphasized later—it fortunately contained next to no mathematics). He reread it several times; he did not have to enter school until after Easter.

One of Bavink's statements, however, repelled Carl Adam, namely that our Mind and Feelings, especially our sense of freedom and self-awareness, were not objects of serious science, not objective, but accessible only by subjective introspection—in short, illusions ("epiphenomena").

Carl Adam was puzzled: Had not Freud offered much more than introspection? Had the great philosophers taught in vain? Had Bavink failed to understand Heisenberg?

Carl Adam felt that it was up to him to solve the apparent riddle of self-awareness and free will, the two items immediately accessible to his thinking. He recalls that he took his first scientific decision: to reach clarity on this matter.

According to a later handwritten note, he argued essentially as follows:

"My simple answer was that consciousness did not belong to me alone, and could therefore not be observed by analysis of my brain activity, however precise and scientific. I saw that if I were alone in the world, I could not form a mental image of myself. I would have to take it from others *like* myself. Only those could form a mental image of me, like I could form one of them. I thought that this mutual imaging produces a vast net, extended over all history and encompassing all Men like me. This net was connecting all minds, it was the Mind itself. Of that I was perfectly sure."

"Next, I considered that my image of others was incomplete, only a projection which omitted many details. Observing my image in others, leading to the only possible kind of image of myself, would then be doubly imprecise; and that's the end of it. This insight made me very sad."

He summarized his insight in the following law: "One part of the universe cannot have a precise, complete image in another part."

This line of reasoning is remarkable in that it already contains elements of Hegelian philosophy as well as ideas from the concepts of infinity in modern mathematical set theory.

2.6   Natural Sciences

Carl Adam performed consistently well at school. He had no difficulties learning Latin or Greek. In his junior years, he was a stand-in singer in the world-famous Thomanerchor. But it was his interest and talent in natural sciences that attracted particular attention. One of the teachers managed to obtain access to the National Library in Leipzig for him, where the works of Jewish authors were still available, even after the Nazi book burnings in 1933. Carl Adam was especially fascinated by Albert Einstein's ideas about relativity of time and space, contemporaneity and causality.

Chemistry A friend of Carl Adam's father had been the director and owner of a well endowed private school that the Nazis had forced to close. This friend gave Carl Adam various apparatus and chemicals from the school laboratory. At home he used these to test chemical reactions, grow crystals, as well as pull apart glass rods using a Bunsen burner branched off from the family gas cooker. This enabled him to obtain glass fibers of well defined flexibility, in order to construct his own microgram-precision analytical balance, with which he could work for years.

His school achievements were good—*except in physical education.* In 1938 they were rated "inadequate" (mangelhaft), a year later a little bit better with the grade "sufficient" (ausreichend). For the Nazi-regime, however, physical training was of prime importance, even more so in view of the upcoming war.

In the late summer of 1939, Carl Adam was sent to a holiday camp to improve his physical education. However, it appears that here too he was more interested in *mens sana* than in *corpus sanum*. On the inside cover of a book on organic chemistry he had brought along, he sketched his knowledge and understanding of chemical substances and reactions. As he recalled later, for these sketches he developed a form of graphical representation that was already very similar to the later modern Petri nets.

What concerned him, however, was that he could not, in that graphical language, express his knowledge of inert gases, which do not appear to take part in any chemical reaction. Also later in his life, he would always return to the fundamental question behind this observation: How does chemical catalysis work? Is the catalyst only a side condition for a reaction, or does it actually take part in it, and is then restored afterwards?

It had seemed so simple:

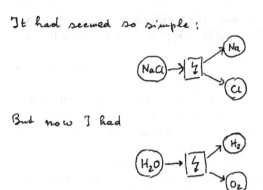

But now I had

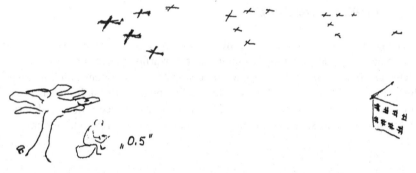

Something was wrong. But this experiment I had done myself, taking the names from the books. I had forgotten to measure Quantity! I quickly wrote 0.5 to one of the arrows, and was saved

I was depressed that War had begun. Less so, that I had used a Fraction.

Fig. 2.4 Handwritten note, probably from the beginning of 2002

Unfortunately, he could not finish his notes for two reasons. Firstly, the space on the book's inside cover was not sufficient. (It is possible that he even smiled a little when he thought about this, because the great Fermat had once had a similar experience.) Secondly, on the first of September the war began. The holiday camp was immediately closed and Carl Adam sent home. Nothing is known of the book's whereabouts. Figure 2.4 shows how Petri later remembered his thought experiment in the holiday camp when the war broke out.

Physics   In 1941 Carl Adam performed what he called his "first scientific experiment, exploiting the deep darkness of war nights for almost a year." He wanted to find out if he could perceive single photons. At his disposal he had only an alarm clock with scintillating hands and dial, which he locked into a drawer. Knowing the size of Planck's constant $\hbar$, and determining the wavelength of the emitted greenish light, he computed how long the clock had to be kept in the dark before the great thing could happen.

Later he recalls the result as follows: "I proved to my complete private satisfaction that my eyes could perceive single photons, but my father and my teachers laughed at me when I told them." On his bicycle he went to Werner Heisenberg's house, who at that time worked at the University of Leipzig, to tell him about the experiment. But Heisenberg was not at home. Unfortunately there was no further opportunity since Heisenberg moved to Berlin soon afterwards.

Forty years later, Petri was engaged as a consultant in the founding of an institute of neurology. (The institute wanted to use computers, Petri had in the meantime made a name for himself in this field.) He asked one of the neurology specialists, if the human eye could possibly discern separate photons. The specialist told him that was indeed the basis of their daily work. As Petri recalls: "I was almost moved to tears, and strongly recommended the foundation."

2.7   Artillery Assistant

At the age of 16, Carl Adam was drafted as artillery assistant (known as *Flakhelfer* in German, i.e. students deployed as anti-aircraft warfare helpers).

This prompted him to reflect on the precision of measurements, and the responsibility involved, because negligent or false adjustment of the sighting instruments could have severe consequences for the gunners. The officers had to determine height, distance and speed of the attacking planes using simple instruments, then

Fig. 2.5 Petri as artillery assistant

compute the target parameters and hand them on to the gunners. But how could the actual operators be held responsible, if their orders were based on inexact observations and conclusions?

In more general terms: What conditions do measuring devices have to satisfy, in order that the users can actually be obliged to assume the responsibility for reading off the values. In fact, this would become one of the leitmotivs of Petri's later research: the development of a comprehensive theory of measurement founded on practice.

2.8   Air Force

In 1944 Carl Adam was drafted to the military; after passing a "Notabitur" (early graduation from secondary school due to war-time requirements). For fear of severe war injuries and a consequent life without dignity, which he observed in many wounded soldiers returning from the battle field, he applied for service as a bomber pilot. His idea was, that within such a service, one could at least expect a clear-cut distinction between life and death. Owing to the shortage of pilots he was in fact accepted, even though he could not demonstrate the political loyalty that usually would have been required.

On April 2, 1944, in a letter to his friend, classmate, and later also brother-in-law, Rudolf Wienhold, that he sent from the training camp in Eger (now Cheb in the Czech Republic) he joked:

Dear Rudolf!
Now I have been an airman for a week, but not you, teehee! I have already finished my recruitment training, and will shortly be promoted. My leave pass has already been issued and is waiting in the orderly office. Knight's cross has already been applied for. A living-dump in Siberia has also already been ordered. (As a precaution, moreover, a job as a dishwasher in USA is arranged.)

Later on he becomes more thoughtful:

When can I start to think of flying? I have not the slightest idea. Maybe already this week, maybe next year. Maybe not at all.

The last possibility turned out to be the correct one. Because of the allied invasion in France on June 6, 1944, he was ordered to the front after all. Very soon he was taken prisoner by British soldiers, and brought to a prisoner of war camp in England. In his short time in the field, there was however *one* war injury he had not managed to avoid: he had become addicted to chain-smoking, and he never found any remedy for it until his death.

2.9   War Captivity

During his 4-year captivity, Petri was treated very decently by the British, for which he always remained most grateful. The camp had a library, provided by anonymous benefactors. He received private lessons from a fellow-prisoner who was a mathematics teacher. He was allowed to work as a land surveyor, and contribute to the planning of a new suburb of Walsall in the West Midlands. In so doing he had to solve challenging problems in surveying methods, for instance the arrangement of concentric ellipses in a hilly terrain, for a bypass road (still clearly visible as *The Oval* with geographic coordinates 52°28'52.86"N, 1°59'13.27"W). This was the first application of his insights on measurement, that he had gained during his time as an artillery assistant.

He was also allowed to exchange letters with his parents. With the pocket money he received, he used to buy shoes, and send them home. In the occupied post-war Germany these could then easily be reconverted into cash.

Presumably in 1948, he was allowed to graduate in the final secondary-school exams in Birmingham, originally intended as second-chance education for British adults.

In December 1948, Petri was released back to Germany.

University, Academe, Family

3

Contents

In 1949 Petri intended to return to normal life and begin his studies at the university. At that time, as a member of an allegedly privileged academic family, he would probably not have been admitted to the University of Leipzig, which was now part of the Soviet occupied zone, later East Germany.

He decided to go to Hannover in West Germany, where he found accommodation with an uncle of his father, Pastor Otto Oehlkers, in the suburb of Hannover-Linden. The pastor's home was already crowded with relatives who had fled from East Germany; but with the welcome words: "You *are* a descendant of the Reformer", Petri was nonetheless accepted. (It was Oehlkers who had traced back the Petri family to the Swedish reformer Olaus Petri.)

3.1   School Graduation: Take 3!

Unfortunately, neither his German war-time exam, nor his English school exam from his time as prisoner of war, was accepted as a certificate of eligibility for a university entrance. He had to pass a third exam. To prepare for this, he attended evening classes. There he had his (as he put it) "first encounter with clear,

© Springer-Verlag Berlin Heidelberg 2015
E. Smith, *Carl Adam Petri*, DOI 10.1007/978-3-662-48093-9_3

Fig. 3.1 Petri with parents around 1950

understandable mathematics." He was tutored by the blind mathematician Helmut Epheser, who at that time was working on his qualifying dissertation (*Habilitation* in German), and later as professor in applied mathematics, in Hannover. After 1 year Petri passed his (this time really the last) exam.

Remarkable is a discourse he gave as part of the examination. In the year 1949 the 200th birthday of Johann Wolfgang von Goethe was celebrated throughout Germany. Both West and East were eager to incorporate Goethe's oeuvre into their own political views of the world. Unaffected by these attempts, Petri took the commemoration as an occasion to question Goethe's scientific investigations. He begins:

> Gentlemen, today I want to give you an account of Goethe's scientific investigations. Actually, I could just as well have named my topic Goethe's shady side, because I would like to propose, that we for the time being, try to liberate ourselves from the prejudice that whatever is written by Goethe must be great and good, simply because it is by Goethe, but rather take a closer look at his scientific—*studies*, a term which I shall prefer here.

Petri wants to shed light on the "peculiar dichotomy between Goethe's outstanding significance for the spiritual world, on one hand, and his almost complete irrelevance in science." He elucidates this by means of the "Farbenlehre" (Theory of Colors), which Goethe himself considered his most important work. Petri recalls the known deficiencies, and refers in detail to Goethe's dispute with Newton. He arrives at the conclusion that Goethe is in fact not looking for understanding, but rather for "harmony". True to his rational view of the world, Petri firmly takes the side of Newton, declaring that the true harmony, which Goethe claims to aim at, is already intrinsic to the elegant mathematical description.

The final part once more testifies how Petri, already in his young years, is disposed to question even widely established thought constructs and authorities, something which would be influential throughout all his later work:

> His [Goethe's] errors arise from the fact that he rejected everything that was there beforehand, an approach that no serious scientist can afford to take. Today we have realized that the fields he excelled in, were not nature and insight, but rather the relationship between nature and art.
>
> In a word, the Farbenlehre is physically meaningless, but significant from an artistic point of view. Goethe himself has shown us that philosophically he did not reach beyond the Ancient Ones, i.e., Aristotle, with his science.
>
> In conclusion, I suggest that we interpret his scientific works, like all ephemeral efforts, only as parables, as parables referring to another reality, which today we call psychological. Indeed this is not so far-fetched, if we consider that in his youth he was deeply involved in mysticism, astrology and the other so-called Paranormal Sciences. And if we then additionally turn a blind eye to his errors, then we arrive at a harmonic Goethe-view, as the phrase goes, but moreover, and much more importantly, at a straight and uncomplicated understanding of our great poet.

3.2   Student Years

He received his certificate, and in 1950 he was finally allowed to begin his studies at the Technische Hochschule Hannover (Technical University of Hannover, today Leibniz University). But which discipline? Until now he had always favored chemistry. However, because of his mathematical tutor Epheser's influence, his interest in mathematics had now been awakened. He visited both departments and, literally, followed his nose. In the chemical institute there was a ubiquitous smell of chemicals. He did not remember any such unpleasant odors from his home laboratory. He decided to study mathematics.

During his studies he took up a job at the Niedersächsische Landesamt für Statistik (a regional State Office for statistics), and also worked as tutorial assistant at the university. He was asked to hold lessons in financial mathematics and practical mathematics, and also to teach geometry for engineers and architects. He recalls: "I was thus for the first time compelled to articulate my insights into measurements, and to share my experiences with others."

In the second part of his course from 1951 onwards, he received a scholarship from the Studienstiftung des deutschen Volkes (German National Academic Foundation, sometimes referred to as "Germany's secret elite university".)

In August 1956 he passed the graduation exam for Diplom-Mathematiker. The topic of his diploma thesis was *Application of mean-value methods for the numerical solution of eigenvalue-problems in differential equations*. The thesis is an example of solid, if not revolutionary, mathematical craftsmanship.

3.3   Hannover Computer Center

The mark he received in his exam was sufficiently good for his teacher, Prof. Heinz Unger, to offer him a temporary assistant post. Part of Petri's duties consisted in the maintenance and support of the University Computer Center. Already in 1955 he had been sent to the German branch office of IBM for some months, in order to get acquainted with practical applications of data processing on an IBM 650 computer.

At the instigation of Petri and one of his colleagues, Jürgen Esch, a machine of this type was acquired for the center; at that time rather unusual, because most German universities were equipped with Zuse Z22 computers. They believed, however, that it would be easier to interact with other international development work using an IBM computer. There was, moreover, a significantly greater range of available programs for the IBM machine. Petri wrote various sophisticated "Tabellenhilfsprogramme" ("table support-programs"), small routines for example to start the machine, each fitting on a single punch card.

Floating-Point Arithmetic   In 1957 Petri wrote his own program for the computation of floating-point arithmetic for the IBM 650. Floating-point numbers are used to represent rational numbers in a computer. Mathematically, the rational numbers comprise a so-called *field*, for which a set of useful rules can be formulated. For instance, in every field the *associative law* $a+(b+c) = (a+b)+c$ holds. In contrast, in *floating-point arithmetic* this is *not* necessarily true, since a machine only permits finite representations. In a 7-digit computation with the numbers $a = 1234.567$, $b = 45.67834$, $c = 0.0004$, we get for instance $(a + b) + c = 1280.245 \neq 1280.246 = a + (b + c)$, owing to rounding errors.

This observation prompted Petri to strive for a self-contained finitary discrete mathematics, based directly on the given finite values and their manipulations, instead of interpreting them as imperfect substitutes for "real" infinite-precision numbers. Clearly this idea is closely related to his approach to practical measurement techniques, mentioned above. In fact, the idea of a finitary mathematics would always remain a prominent idea throughout his work.

Formal Pragmatics   While working in the computer center Petri had another insight, which would prove to be crucial in his subsequent ideas. He noted that magnetic tapes and punched cards had a special common property: They are carriers of *information*. When this information has been used, the carriers lose their status, although they have not changed physically.

He became interested in the pragmatic aspect of information, which later grew into one of his leitmotivs: the development of a *formal pragmatics*. Pragmatics is here not to be understood in its colloquial meaning "following practical rather than theoretical considerations", but as the level above semantics and syntax in the science of general semiotics. Besides the grammatical structure of communication (syntax) and the meaning of the language symbols (semantics), pragmatics is also concerned with the *relationships* between symbols and their *effect* on the *involved personal actors*.

Petri often maintained that it was in this context that he arrived at the graphical notation of his nets. The circles reflected the form of the tapes, the squares the form of a computing machine. A token on an input-circle meant that the information on the tape had not yet been processed, a token on an output-place, that it had. The transition thus described the change of the information's pragmatic status.

As a last task at Hannover, in the winter term of 1958/59 Petri gave a course on programming the IBM 650.

3.4 Family

At Christmas 1955 Petri married the girlfriend from his schooldays, Christel Wienhold from Leipzig. She was the sister of his class mate Rudolf Wienhold, already mentioned above. Petri had known her since he was 10 years old.

When the two were still engaged, she worked as a foreign-language correspondent and interpreter in Braunschweig, about 70 km (45 miles) from Hannover. She lived in a girls' home; men were not allowed to enter the premises. When the two wanted to go out together, Petri had to wait for her outside the front door. Because there was no doorbell, they had agreed on a whistled tune, which the other girls could not imitate, although they tried time and again. They used the theme of Beethoven's canon *O Tobias!*—for the untrained ear not easily distinguishable from the well-known "cuckoo-cuckoo"-call. The musical education at the Thomasschule had paid off.

It is not known whether the piece in Fig. 3.2 dates from this period, but it certainly corresponds to Petri's spirit at that time. (Another possibility is that it refers to the first living being in space, the Soviet dog Laika 1957.) Translated into English it says: "Science has discovered why the dog barks at the moon: It only mimicked the moon itself who had begun the yelping."

Text und Musik: Carl Adam Petri

Die Wis-sen-schaft hat fest-ge-stellt, war- um der Hund den Mond an-bellt : Er

hat den Mond nur nach- ge- äfft, der hat zu- erst ge- kläfft.

Fig. 3.2 Composition

At times the written communications between the two assumed rather curious forms. Possibly in order to save postal charges, Petri sent his fiancée letters from the university, written on the chair's official writing paper. He disguised his actual messages in apparent answers to supposed mathematical requests. Moreover, this game of 'hide and seek' and persiflage of contrived bureaucratic terms certainly gave them great pleasure. One such letter—dated June 9, 1955—still exists. Apparently referring to a mathematical question not explained in detail, it begins:

Dear Miss Wienhold!
The chair regrets that it is not able to respond to your most highly appreciated inquiry to the full extent it deserves. We must ask you to be willing to take into account that the investigations into the problem area you touch upon are exceedingly difficult and still in a state of flux. Moreover, you should not expect too much from the mathematicians' endeavors to raise this veil—more precisely, to raise at least a group of corners of this veil-system, even though, as we can assure you, these endeavors amount to a desperate effort. Presently various outstanding scientists—in particular at our chair—devote themselves to these tasks, having received—which is very gratifying—valuable suggestions from groups of laymen.—We therefore have to ask you to regard the following deliberations by our specialists as a preliminary result, with all due reservations with regard to further scientific progress. They do however—as we can assure you in all confidence—faithfully reflect the present state of the research.
We do not request any remuneration for our efforts, but nonetheless cherish the hope for your goodwill on appropriate occasions.

Fig. 3.3 Left marriage 1955, *to the right*: Petri with son Tobias

After some general remarks on mathematical game theory, the letter ends in a manner not immediately comprehensible to outsiders:

> Moreover, players have become known, who prefer the game of chess or certain other games; albeit only because in these games any involvement—also indirect—of third parties can be excluded. This probably is the case only with asocial and misanthropic characters. With this assessment we are however in danger of transcending our scientific competence. Nonetheless we hope to have served you with our declarations, and remain
>
> your faithfully devoted Chair for Descriptive Geometry and Practical Mathematics. Technical University Hannover.

Two years later, in 1957, their only child was born. It was a boy, to whom the parents—in remembrance of their identifying signal—gave the name Tobias. It was only afterwards that they discovered that the name in fact occurred amongst their ancestors. Petri's great-great-grandfather Tobias Petri Sr. (around 1790–1860) was a farmer in Eggersdorf, today part of the municipality Bördeland in Saxony-Anhalt.

In the child's education, methods of computer science were used already at an early stage. Figure 3.4 shows a pedagogical game designed by Petri. As can be seen, at that time he still used flow charts rather than nets.

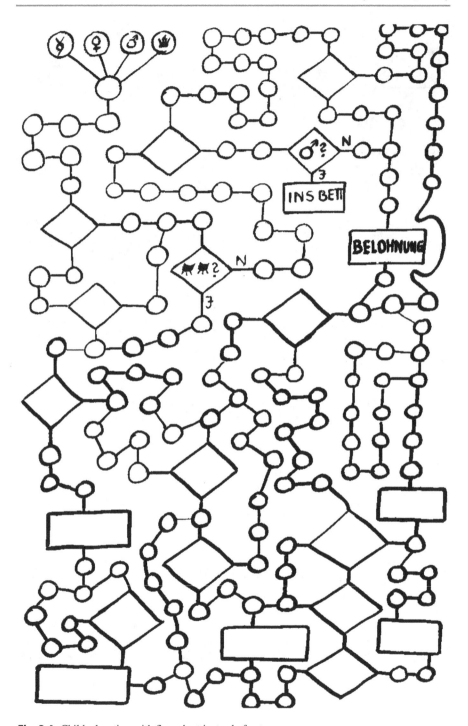

Fig. 3.4 Child education with flow chart instead of net

3.5   Bonn

In 1958 Petri followed Prof. Unger to the University of Bonn as his scientific assistant. After arriving in Bonn, he shifted his professional focus once and for all to, what many still disdainfully called "electrical mathematics": He gave theoretically oriented courses on, among others, the structure and programming of electronic computers, and formal languages, but also practical courses on the programming languages Fortran and Algol. From that time there still exists a copy of Petri's notes for a lecture on *Formal Languages and their Translation*, in which he explains the general principles of machine-based language translation, using Algol 60 as an example.

Standing between these poles of theory and practice, he became aware of a certain discrepancy between the *use* of real computers, on one hand, and their mathematical *foundations*, on the other. This observation led to the topic of his doctoral thesis.

Thesis on Automata

4

Contents

Petri obtained his doctorate in 1962 at the Technical University of Darmstadt, with the thesis *Kommunikation mit Automaten* (Communication with Automata) that he had submitted the year before.

The first examiner was Prof. Alwin Walther, head of the institute for applied mathematics in Darmstadt. He was one of the pioneers of machine-computing technologies in Germany, and at the time president of the International Federation for Information Processing (IFIP). The second examiner, Petri's teacher Prof. Unger, had himself also been one of Walther's students. The final oral examination took place on June 20, 1962.

As Petri recalls: "Walther asked me about the ambiguity of the title of the thesis: did I mean that automata were *partners*, or else *tools* in communication? I replied that in the thesis, I would have to have used Turing's image of Artificial Intelligence, to comply with the current philosophy, *but* that I was certain that the machines of information technology would soon be developed to realistically and usefully serve as much more than tools in communication. I had used the ambiguity quite intentionally, and was happy that it had been noticed."

Occasionally, even a third possible interpretation of the title is suggested, namely that it might also refer to communication *between humans* by means of automata— in the sense of an anticipation of computer networks. However, there is no evidence supporting this suggestion.

He received his doctoral degree "summa cum laude". The thesis won an award for the best in the academic year 1961/62. It was published as No. 2 in the series *Schriften des rheinisch-westfälischen Instituts für instrumentelle Mathematik* (Publications of the Rhenish-Westphalian Institute for Instrumental Mathematics)

© Springer-Verlag Berlin Heidelberg 2015
E. Smith, *Carl Adam Petri*, DOI 10.1007/978-3-662-48093-9_4

Fig. 4.1 Petri around 1963. Satisfied with his thesis?

at the University of Bonn. (No. 1 in that publication series was a work by Fritz Krückeberg on numerical integration of ordinary differential equations. Krückeberg later became executive director of the GMD and one of Petri's colleagues as institute director.)

4.1 Communication with Automata

In the thesis Petri discusses the shortcomings of conventional automata theory for the description of large computing systems. He proposes an approach that strictly adheres to the underlying physical realities. The crucial decision is to base the description model on *local* operations, interacting with each other *asynchronously*, so as to avoid the assumption of a global timing common to the whole system. This idea clearly reflects the insights Petri had gained in his studies of Einstein's relativity theory.

By *automaton* he understands, as usual in theoretical computer science, the abstract model of a computing machine. By *communication* Petri means "all manifestations of information flow", here more specifically the input/output-behavior, that consists in supplying the machine with sequences of signals, which it evaluates and then reacts with corresponding signal sequences as output.

Iteratively and Recursively Declared Expressions Every concrete computing device actually corresponds to a *finite* automaton, that can only distinguish between a *limited* number of input sequences. A finite automaton is by construction only able to recognize sets of sequences, characterized by so-called *regular expressions*, or—in Petri's terms—that are *declared iteratively*. However, in general more complex, *recursive* expressions are required for communication with automata, for instance to check whether an arithmetical term is correctly bracketed, or in the description of compilers translating high-level programming languages into machine-executable instructions.

Petri illustrates the difference between iteratively and recursively declared symbol strings by means of the sets of all sequences

(1) $1 \ldots 101 \ldots 1$, consisting of an arbitrary number of 1s followed by a single 0, then again followed by an arbitrary number of 1s,

and, respectively, of all sequences of the form

(2) $1 \ldots 101 \ldots 1$, where now however the suffix contains *the same number* of 1s as the prefix.

Using the iteration operator $*$ to denote 'zero or more' repetitions, the sequences of type (1) can be characterized by the regular expression $1*01*$. It is also straightforward to define a finite automaton, which precisely accepts input sequences of this type: Such an automaton must only be able to assume two different states, a *start state s* and an *accepting state a*. When the automaton in its start state s is supplied with the input string 1011, it reads the first symbol 1, and remains in the state s. Now it reads the symbol 0, and switches to the state a. In this state it reads the next 1 and remains in the state a. Finally it reads the last symbol occurrence 1, and by remaining in the accepting state a indicates that the input string has been accepted.

The situation is very different in strings of type (2). When the sequence is sufficiently long, a given automaton cannot record the number of 1s in the prefix with the help of its finite number of states, which in particular means that also a length comparison with the second 1-sequence becomes impossible. Consequently, there can also be *no* regular expression that characterizes precisely all sequences of type (2).

However, sequences of type (2) *can* be generated by a so-called *recursive* rule system, consisting e.g. of the two rules $S \to 1S1$ and $S \to 0$. Beginning with the start symbol S, first the sequence $1S1$ is generated, in which then in turn S is replaced by $1S1$, yielding $11S11$. This procedure can now be repeated arbitrarily often, until the production is terminated with one final application of the second rule $S \to 0$.

Obviously every sequence of type (2) is also of type (1), but the more *restrictive specification* in (2) cannot be formulated by a regular expression. A method of defining expressions is stronger if it permits more *precise distinctions*, not if it can generate bigger sets.

Now, it is clear that a given automaton *can* doubtlessly process recursive expressions *up to a certain size*. When required, it may then be extended to deal with larger expressions. This is analogous to, for instance, the transition to computers with a larger address space, from 8 bit to 16, to 32 and currently to 64 bit.

Petri's Postulates The finiteness of real machines was usually considered a *practical* limitation, and, as Petri critically remarks, "not as something that should by itself also be part of the theory." In his opinion an adequate approach can only consist of the recognition of the governing physical laws as *foundational conditions* for automata, and include them into the theory. He specifies his requirement in the form of a manifesto as follows:

We assume the following postulates:

1. There exists an upper bound on the speed of signals.
   There exists an upper bound on the density with which information can be stored.
2. Automata of fixed, finite size can recognize, at best, only iteratively declared classes of input sequences.
3. Recursively declared classes ... can be recognized only by automata of unbounded size.
4. In order for an automaton to solve a ... recursive problem, the possibility must be granted that it can be extended unboundedly in whatever way might be required.
5. Automata formulated in accordance with automata theory will, after a finite number of extensions, conflict with at least one of the postulates above.

The fact that signal-propagation time increases with the size of a system, and therefore requires special considerations, was for instance already recognized in the design of one of the first "supercomputers", the Cray-1 from the year 1976. A horseshoe form of construction was chosen, which allowed reduced distances between components, thus reducing propagation time.

But the reduction of signal-propagation time by increasing information density is also limited by physical laws. This problem was not unknown to theorists. Petri refers to proposals by the computer pioneer John von Neumann to supplement the automata model with *delay elements* to accommodate for the speed deceleration resulting from system extensions. These approaches were not very successful, however, since it soon became clear that, along with physical system growth, the control by means of a global clock generator becomes increasingly more difficult; in other words: the assumption of a global system time becomes more and more untenable. On the contrary: in large systems, maintenance of simultaneity is only possible through active synchronization. Synchrony can only be achieved through communication.

A typical example, nowadays familiar to everyone, is the *global positioning system* GPS, used e.g. in car navigation. The distance to the satellites is determined via the propagation time of radio signals. This however requires a very tight synchronism between the clocks in sender and receiver, which has to be reestablished regularly.

According to Petri, in large systems we have to abandon the concept of a global system state at a well-defined point in time t. A first consequence of his postulates is that a realistic model of computing systems must be based on *asynchronous* operations.

Asynchronous Switching Networks In the central part of his thesis, Petri works out the specification of suitable switching elements, and shows how to connect them into asynchronous switching networks. The formal development is of a very technical nature, and does not yet exhibit the later elegance of typical Petri nets. It is therefore not surprising that the thesis is often cited but seldom read. Petri himself points to the "arbitrariness of the constructions, ... which are by no means real suggestions for building switching networks, but rather preliminary tools used for proofs." Elsewhere he later speaks of the "earliest clumsy constructions" used in the thesis [9].

The basic switching elements are chosen such that both they themselves, as well as the networks built with them, are physically realizable in the sense of the above postulates. In particular, the behavior is not to depend on temporal assumptions.

Petri illustrates the idea with the help of the classical *AND gate*, where logical values are represented by electrical voltage. An AND gate has two inputs and one output. When both inputs have the value 1—represented by, say, a voltage of 5 volts—then the output value should also be 1, i.e., the output voltage also reach 5 volts.

Naturally this takes some time. The network system must already from the start be designed so as to permit a signal to reach a stable state within one clock cycle.

In contrast, Petri only requires that a switching element may operate *as soon as all input conditions are satisfied*, without the need for any temporal assumptions.

Another distinction from classical switching is that the proposed elements *transform* input- into output-conditions, i.e., the inputs are "consumed", and are afterwards *no longer available*. To Petri this assumption is of crucial importance, because it is only in this way that physical realizability at *all levels* can be guaranteed. He recalls that, for instance on the level of quantum mechanics, already an arbitrary observation is accompanied by a change of state. Another root for this requirement is obviously to be found in the considerations on chemical catalysts, that he made during his school years. On a higher level, this type of "destructive reading" was for instance known from the magnetic core memory used in computers up to the 1970s.

For the formal description of his switching networks, he extends the language of regular expressions by a *parallel operator*, in order to specify independence between different parts of the network. This approach is nowadays common in various theoretical models of process algebra, but is also used in modern programming languages for the specification of so-called *asynchronous threads*.

As a matter of fact, in a conference on the development of the programming language Algol 60 in the year 1959, Petri had already proposed to include a parallel operator into the language, but could not as yet win through with the idea.

OR Gate Conventionally, switching circuits are assumed to be built from certain logical gate-types, such as the already mentioned AND gate. As a first application of his approach, Petri shows how another important gate can be represented within his framework: the *OR gate*.

It is a remarkable fact, that the treatment of the OR gate is substantially more difficult than for instance that of the AND gate. An OR statement is true when *one* of the preconditions hold, but also when *both* hold. At a fundamental level, a circuit that realizes such a gate must thus permit *alternative* actions. In certain cases therefore a *conflict* may arise: When two actions with shared inputs are enabled, it is not clear, which one will occur. Petri develops a complicated construction, intended to avoid such conflicts. Figure 4.2 shows an outline. The internal structure of the modules NA, NB and NC is not depicted, but likewise not immediately comprehensible.

In the light of more recent insights, it is not impossible that a subtle error may have found its way into the construction. More on that in Sect. 6.5.

The Asynchronous Turing Machine A large part of the thesis is devoted to the demonstration that the asynchronous approach has the same computing power as conventional models.

Specifically, Petri shows how *Turing machines* can be realized by his switching networks, in order to prove "that the theory of synchronous automata is contained in a more generally theory of communication forms." Reformulated in modern net terms, he shows how the behavior of Turing machines can be reproduced by means of condition-event systems of rank 2, i.e., where every transition has a most two input- respectively output-conditions.

The Turing machine is a reference model for computability, in the sense that everything that is at all mechanically computable, is already computable by such a device. If you can prove that a computational model can emulate a Turing machine, it inherits, so to speak, the full computing power.

A Turing machine possesses a *control unit*, which can assume one of a finite number of states, and an infinite tape divided into cells. Each cell can either be empty or marked with a symbol 1. At every stage, exactly one cell, the *current cell*, is scanned by a read/write-head. The machine's behavior is determined by an internal program; depending on the machine state and the content of the current cell, the machine can change the content, move the head to an adjacent cell, and shift into a new state (see Fig. 4.3).

Petri begins by developing a net representation of the control unit and the read/write-head, and shows how every computation of a Turing machine on a *given finite part of the tape* can be realized asynchronously by nets.

Memory Extension In compliance with his postulate that an automaton must on demand be indefinitely extendible, Petri formulates rules for the request of additional memory, when the machine reaches the end of the finite tape. These rules are again expressed in terms of nets, more precisely within a "logical plan" specified within the net language "according to which the additional memory has to be provided by a meta-machine (or the user)." But also this "meta-communication" should as far as possible adhere to strict formal rules.

A corresponding rule set was then later specified and worked out by Hartman Genrich, one of Petri's closest collaborators throughout the years, see for instance

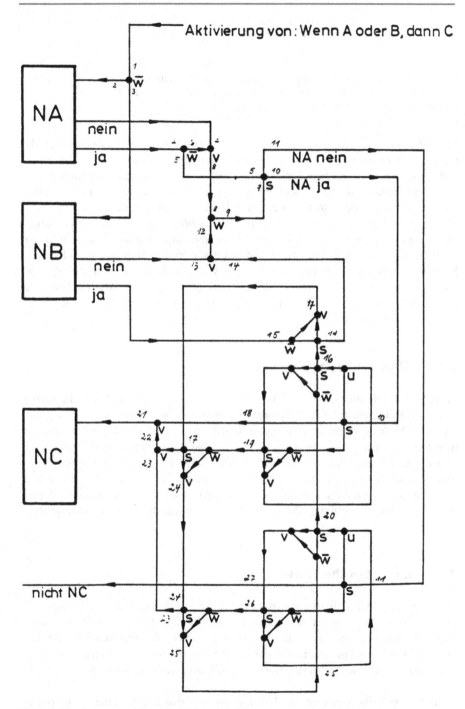

Fig. 4.2 Petri's representation of an OR gate in [4]

Fig. 4.3 In the state q the Turing machine reads the symbol a in the current cell (*left*), replaces it by a', moves the head to the right and shifts to the next state q' (*to the right*). The letters $a, \ldots, d, a'$ denote respectively either an empty cell or the symbol 1

Ein Kalkül des Planens und Handelns (A Calculus of Planning and Action) [3]. (In the Petri puzzle mentioned in Chap. 1, this extendibility-idea is reflected in a rule that the size of the game board may be modified during the course of a game.)

This final part of the thesis definitely proceeds beyond automata theory in a strict sense, and rather points to the later "high-level nets" of general net theory. Petri observes: "When we speak of communication *with* a net, we imagine ourselves in the position of a partial net, but we do not immediately have the possibility of applying to this partial net the same forms of description as were arranged for nets. Rather we must use a form of communication with automata, different from the one that consists in simply feeding the machine with input sequences. Also the language used will in general be a different one, regardless of whether the input is declared iteratively or recursively."

4.2   Alternative Approaches

In the early preliminary work for the thesis, Petri also investigated two alternative system models, which he called *assignment systems* and *loosely coupled systems*. Again in both of these approaches, it is the concept of asynchronicity that is in the center of attention. In assignment systems, the central notion is the *value assignment* to variables, as in classical imperative programming languages. In loosely coupled systems, it is the concept of mutual exclusive access to shared resources that forms the starting point. The loosely coupled systems were developed further by co-workers in the GMD; but ultimately it was the more versatile general net approach that prevailed.

4.3   Impact of the Thesis

In its original intent, Petri's thesis is concerned with the design of computing models, i.e., with computer architecture. However, nets did not at first have their impact in that field, but rather in such systems, where the distributed nature was evident: in industrial production processes, in telecommunication, in traffic control, later also in computer networks. Basically, this situation continued up through the 1970s.

In the 1980s the idea arose to design computers that followed the model of the human brain. It was based on the assumption that the power of the brain arises from a complex interaction of an enormous number of neurons, each of which,

Fig. 4.4 Petri around 1963. Thinker and musician

viewed in isolation, exhibits a rather simple behavior. This gave rise to the approach of so-called *massively parallel systems*, where massively many simple processing elements were to be connected to one another. It was in these orders of magnitude that the physical limits described by Petri came to be increasingly relevant.

One of the best known examples of such a massively parallel machine, is the *Connection Machine*, developed by the researcher Daniel Hillis at the Massachusetts Institute of Technology (MIT). Its first incarnation, the CM-1, consisted of 65,536 processing elements, each of which was extremely simple: only capable of handling 1 bit. It was built by the Thinking Machines company, founded by Hillis, from 1983 onwards. One of these machines found its way to the GMD.

During the development, and in particular during the construction, Hillis directly experienced the significance of the physical limitations. He observes: "We are discovering that it previously appeared as if we could connect a wire to as many places as we wanted, only because we did not yet want to connect to many places. We have been forced to notice that we cannot measure a signal without disturbing it; for example we must drive a wire with power proportional to the number of inputs that sense it."

Today Petri nets have also gained widespread use in the description of computer architectures. But naturally, they prove themselves to be particularly useful in contexts where coordination of asynchronous actions is of prime importance.

The Maturing Years

5

Contents

In the following years, Petri generalizes the insights from his thesis, always closely guided by the requirements of physical compatibility. The ultimate goal he strives for, is to find a common denominator for automata theory, switching circuits, geometry, and physics.

5.1   Asynchronous Information Flow

In the year between submission and acceptance of his thesis in 1962, Petri already began with the further development of his approach. In the same year, 1962, at the end of August, he presents a paper at the IFIP-conference in Munich, titled *Fundamentals of a Theory of Asynchronous Information Flow* [6]. Here he puts the focus on the characterization on the *processes* that can run in a distributed system, i.e., physical processes that can serve as carriers of information processing. As a fundamental concept, he identifies so-called *signal streams*. Signals are here understood as causes and effects of state changes, or, expressed differently: A distribution of signals comprises a state, events change that state.

As he explains in more detail later in *Fundamentals of the Representation of Discrete Processes* [8], the physical concept of "time reversal invariance on a small scale" is of crucial importance to him. Based on that concept, he argues that for an adequate theory of signal streams, it is sufficient to consider events that have at least *two* causes and *two* effects.

© Springer-Verlag Berlin Heidelberg 2015
E. Smith, *Carl Adam Petri*, DOI 10.1007/978-3-662-48093-9_5

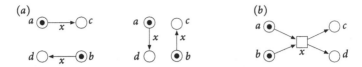

Fig. 5.1 (a) Two representations of a T-element x in [6], (b) modern net form

A coupling of such signal streams leads to the notion of *process*. In [6] Petri formulates his approach as follows: "We regard every process as composed of discrete events T. To describe relations between events we use the concept of a state, complementary to that of an event. We shall say that an elementary event alters the state of a finite number of entities called state elements S."

S-elements as carriers of signals already occur in the later parts of the thesis, including the graphical representation by circles, with tokens to indicate the presence of signals. The nowadays customary representation of T-elements has however yet to arrive. In [6] we still find: "A T-element is shown as a pair of directed arcs connecting two pairs of S, the arcs of any one T bearing the same label. ... Every T-element may thus be represented in two different ways." (See the two representations in Fig. 5.1a.)

This representation is somewhat astonishing, since Petri actually had used the square-notation already on two occasions: in the description of chemical reactions in his school days, and then again to denote the transformation of the pragmatic status of tape storage, during his time at the Hannover Computer Center. It is not until 3 years later, in [8], that the graphical representation of nets as bipartite graphs consisting of circles and squares is (re?)introduced.

The IFIP paper [6] is actually only a rather short note, referring to a future article in the "Schriftenreihe des Bonner Instituts für Instrumentelle Mathematik," where also Petri's thesis had been published. However, no full paper ever appeared.

5.2   The Bonn Computer Center

Petri's further development of his theory unfortunately slowed down for some time. In 1963 he was appointed Kustos (curator) at the University of Bonn, that is "a scientific staff member entrusted with administrative tasks in a university-institute or -seminar". Since it was an important official appointment, he had to be sworn in. According to protocol, he was asked whether he wanted to take the oath with or without the formula "So help me God". Petri is said to have responded: "I swear without *any* reservation whatsoever."

As before in Hannover, Petri began with the establishment and management of the University Computer Center, this time however on a significantly larger scale. In spite of his colleagues' substantial concerns ("Much too large, will never be used to the full capacity") he pushed through the acquisition of two IBM computers, a mainframe 7090 machine, and a smaller satellite IBM computer 1410, intended to reduce the main machine's workload by taking care of data-input/output and

-management. Petri rewrote a substantial part of the operating system *Fortran Monitor System*, in order, for instance, to ensure the cooperation of the two machines via the shared tape units.

The Bonn Computer Center became the second largest non-commercial center in Germany, second only to the "Deutsche Rechenzentrum" (DRZ) in Darmstadt. The computing resources of the 7090 machine were made available to all of the university's faculties, and the machine was soon used to its full capacity. The physicists could evaluate bubble chamber experiments and simulate electron scattering processes, the philosophers compiled an index of the works of Immanuel Kant, and linguists tried out the automatic translation of natural languages.

But Petri also took care of the more down-to-earth issues of computer operations: He himself swept the floor of the computer room, because, as he recalls: "The cleaning ladies did not do it sensibly enough."

In 1966 Petri was promoted to Oberkustos (senior curator). In that same year he established a research group on the foundational problems in data processing, which then 2 years later was incorporated into the GMD.

5.3 Israel

Besides his work in the computer center, Petri made a—as he said—"short excursion" into phonetics and linguistics, and their processing in the computer. In this way he came into contact with Victor Yngve, author of the first programming language for string processing, which he developed for IBM machines of type 700/7000 in the early 1960s, together with co-workers at the Massachusetts Institute of Technology (MIT). Yngve designed the language for research in, e.g., the fields of machine translation and natural-language speech input.

Yngve must have been impressed by Petri; he told the Israeli philosopher and mathematician Yehoshua Bar-Hillel about him and his PhD thesis. Bar-Hillel is best known for his pioneering work in computer translation and formal linguistics. So it came about that Petri was invited to the 1964 *International Congress on Logic and Philosophy in Science* in Jerusalem. Bar-Hillel was also able to obtain a grant to finance Petri's stay, in spite of the scarce resources of Israel.

It was also Bar-Hillel who recognized that Petri's work reached far beyond automata theory. As Petri recalls later, full of gratitude: "*He* told me to carry on. *He* told me that pragmatics has priority over logic. He introduced me to the present philosophers. He showed me the Qumran scrolls. Laying aside his heavy duties as organizer of the congress, he gave me private talks. He treated me as though I were a very important person. He told me not to be intimidated by philosophers. I promised. Next day, I *was* intimidated by McCarthy, Kleene, Rabin and Zemanek."

John McCarthy was a US American logician and computer scientist. He was the inventor of the programming language LISP. Stephen Kleene is considered to be one of the founders of theoretical computer science, in particular formal languages and automata theory. Michael Rabin is an Israeli computer scientist. He has made

Fig. 5.2 Petri in Israel in
1964

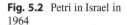

Fig. 5.2 Petri in Israel in
1964

valuable contributions in the field of cryptography based on prime numbers, as well
as in automata theory. Heinz Zemanek was an Austrian computer pioneer.

5.4   America

At the conference in Israel, Petri came to know Anatol Holt, who invited him to the
USA for the first time. Holt had begun his career in 1952 as a programmer of the
UNIVAC I in a working team led by John W. Mauchly. (The UNIVAC developed by
Mauchly and John Presper Eckert was the first commercially available computer in
the United States.)

From 1963 to 1974 Holt headed a research project on the theory of information
systems, supported by the research agency of the US Department of Defense, ARPA
(Advanced Research Projects Agency). Petri-net research became a central part of
the project, such that Petri and Holt worked intensively together on the development
of nets, first in the USA, and then later also at the GMD.

The two were united in a long but also difficult friendship. Holt immediately
recognized the possibilities of nets, and pushed Petri in their further development.
Holt is said to be the one who introduced the term "Petri net", in a conference around
1967. (However, it is possible that the term was coined even earlier: Tom DeMarco,
inventor of structured analysis, a method to generate formal system descriptions in
the framework of software development, is reported to have heard it already in 1964,
during a project presentation at the Bell Telephone Laboratories.)

In a certain sense, Petri and Holt were however also rivals; Holt later often
maintained that many of the central ideas in nets were originally proposed by him.

The ARPA-Net Beginning in 1962, the development of the ARPA-network, the
predecessor of the Internet, was initiated by the US Department of Defense, and the

Fig. 5.3 Holt and Petri, at Petri's place (*to the left*), 1967 in Scheveningen, Holland (*to the right*)

Massachusetts Institute of Technology (MIT). The objective was to investigate the possibilities of large scale interconnection of computers.

In Boston, Petri gave a lecture to developers of the ARPA-net and MIT researchers, in which he discussed the limits of semantic analysis methods for communication. He recalls: "I told my audience … that, in my opinion, semantic analysis could never come to an end, except in an utterly restricted area: that of writing 'correct' compilers for programming languages which were badly designed. I urged that a radically wider approach was necessary to stabilize the present and future computer-based communication networks: the design of formal pragmatics."

In the same spirit, Petri later often aphoristically characterized semantics as "the infinitely thin layer between syntax and pragmatics."

During its further development, Petri came into contact with engineers and researchers, who investigated difficulties of the ARPA-net that had in the meantime been discovered. Petri could identify various classes of problems, a classification that he later would elaborate in the framework of his *communication disciplines*. More about this in Chap. 9.

Government Advisor At the end of the 1960s, Petri was for a time commissioned as advisor to the US government. With the extra remuneration, he was able to acquire his first own new car, a NSU Ro 80, equipped with a—at that time groundbreaking—Wankel rotary engine. Unfortunately, the revolutionary Wankel technology encountered so many practical problems that production was terminated after some years. Malicious gossip, critical to the future of Petri nets, had it that the Ro 80 was precisely the right car for Petri; just like the nets it was based on a brilliant idea, but useless in practice. Petri drove the car until he turned it in for a new one of the same type in 1976, which he then sold to an enthusiast in 1992. As for nets, there is still no end in sight.

5.5 Physical Determinism

For a deeper understanding of Petri's work, it is absolutely necessary to include his cosmological-philosophical conception of the world. He was a convinced adherent of physical determinism, advocated for instance by Pierre-Simon Laplace, for whom the world is completely determined by initial conditions and its causal laws of motion, such that a demon with a total knowledge could mathematically predict the future of the world.

Also in this respect, Petri was clearly influenced by Albert Einstein, who summarized his determinism in the famous statement: "God does not play dice" (in response to which, incidentally, Niels Bohr is said to have begged Einstein please not to tell God how to behave).

Petri's Universe: Conflict-Free and Informationally Closed In his PhD thesis, Petri had remarked that events may be in conflict with one another, but had also proposed constructions to avoid such situations. His first detailed account on the matter can be traced back to a lecture he gave at the *3. Colloquium über Automatentheorie* 1965 in Hannover [7, 8]. There he maintains that a conflict may only appear when the view is *restricted to a partial net* of a postulated *global* net, which provides all the necessary conditions for the resolution of the conflict. On the border between the partial net and the environment, the influence appears as a provision of additional information. In Petri's own words: "In the global net the information-stream carried by a signal-stream is 'divergence-free', i.e., deterministic. Conflicts are decided only by information-*supply*."

The same idea appears over and over again, for instance in *Concepts of Net Theory* [9], and later in *Interpretations of Net Theory* [10] like this: "A choice between alternatives can arise only when a part of the world—e.g., a system— is considered separately from the rest of the world, e.g., the environment of the system." Also in *General Net Theory* [13] he emphasizes: "If a transition set contains a conflict, then the system described by it possesses a non-empty environment." In *State-Transition Structures in Physics and in Computation* [18] we find: "Deciding a conflict defines one bit of information. From where?" Ultimately from the surrounding universe: "A detailed ... description of the universe contains no conflict situations."

To summarize, Petri maintains that the information to decide a conflict is already present in the vicinity of the involved transitions. The conflict manifests itself, because only a part of the encompassing conflict-free net is considered, and consequently also only the signal distribution on one side of the borderline is taken into account. Within this distribution, conflicts appear, which however have no reality in the larger net.

With regard to this concept of information, Petri concludes already in his 1965 lecture *Fundamentals of the Representation of Discrete Processes* [7, 8]: "We thus arrive at a purely combinatorial conception of information, as sometimes requested by Kolmogorow." Andrei Nikolajewitsch Kolmogorow (1903–1987) was an important Russian mathematician, whose best known achievement is the axiomatization of probability theory. As late as 1983 he claimed that "Information theory must precede probability theory, and not be based on it."

Randomness and Causality Such a determinism of course also has consequences for the concept of randomness. Randomness can mean that (1) no cause exists, but can also mean, that a cause exists, which however (2) is unknown. In computers, random generators of the latter type are usually used: Actually there is a precise mathematical generation rule, but it is of such a complexity that in practice it amounts to non-knowledge. Occasionally it has been proposed to construct "genuine" random generators, based for example on radioactive decay. However, if the world is deterministic, this approach again falls within variant (2).

If, on the other hand, randomness is actually a phenomenon inherent in nature, then ultimately the idea of causality loses its strict sense, and must be replaced by "correlation with a probability close to 1".

For the underlying structures of his non-sequential processes, Petri originally chose the German term *Kausalnetz* (causal net), and thus once more emphasized his point of view. (The commonly used English translation into *occurrence net* is clearly less expressive.)

Hidden Parameters in Quantum Mechanics Against this background it is also not surprising that Petri was very sympathetic to the thesis that the apparent random processes in quantum mechanics do in fact have a determined cause, which however is hidden from observation. He discussed the question with leading physicists in the field, such as John Stewart Bell, who at that time worked at the European Center for Nuclear Research (CERN) in Switzerland. Bell maintained the opinion that there were no such hidden variables. Petri's view had a better reception from the Dutch physicist and Nobel prize winner Gerard 't Hooft, who argues that quantum mechanics should be based on a deterministic theory.

5.6 Reversible Logic

Independent from the question of how information is to be defined, or how it originates, it is certainly of prime importance to understand how it flows and is processed within systems.

In his doctoral thesis Petri emphasized, by means of the classical AND gate, that the use of switching elements with *two* inputs but only *one* output is *irreversibly* accompanied by a loss of information. On the physical level, the corresponding transformation of two input bits into one output bit leads to a dissipation of the excess energy into the environment in the form of heat. In fact it is known, e.g., from the investigation of reversible processes in biology, that for reasons of energy balance, it is advantageous to base information processing on physically reversible processes, whenever possible.

Already in 1965, in the colloquium on automata theory in Hannover mentioned above, Petri proposes a model, in which all Boolean computations are embedded in reversible transformations, and thus—at least in principle—can be implemented by corresponding energy conserving physical processes. He calls these structures *information flow graphs* [8].

The ideas were later taken up by other researchers, who occasionally put them on the market under their own name. Amongst others they presently play a role in the discussion of the so-called *quantum computers*, an approach, at the time of writing still highly speculative, in which the operations of the computer are directly based on the laws of quantum mechanics.

The Petri Gate The central building block in information flow graphs is the switching element shown in Fig. 5.4. Petri calls it *Pfeilfunktion* (arrow function), more precisely the Pfeilfunktion P_2. Here we shall use the term *Petri gate* instead.

The input signals x, y, z to the left can assume the Boolean values 0 or 1. The values of x, y are passed through unchanged to the outputs. On the signal line marked by the tick, the output is computed from the three inputs as $z \oplus x \cdot y$. Here "·" denotes the normal multiplication $0 \cdot 0 = 0 \cdot 1 = 1 \cdot 0 = 0$ and $1 \cdot 1 = 1$, corresponding to the logical AND. As usual we often write xy instead of $x \cdot y$.

The operator $\oplus$ denotes the logical "Exclusive-Or". An Exclusive-Or statement $x \oplus y$ is true when *exactly one* of the two terms x, y is true. In 0–1 notation we thus get $0 \oplus 0 = 0 = 1 \oplus 1, 0 \oplus 1 = 1 = 1 \oplus 0$. The multiplication "·" has precedence over $\oplus$, so that $z \oplus x \cdot y$ is evaluated in the order $z \oplus (x \cdot y)$. For the input values $x = y = z = 1$ we for example get the output-value $z \oplus xy = 1 \oplus 1 \cdot 1 = 1 \oplus 1 = 0$.

Thus, the effect of the Petri gate is that the value of z is *inverted* if and only if x and y both have the value 1. The crucial property of the gate is that it defines an *invertible* mapping between the input values (x, y, z) and output values (x', y', z'). If, say, $x' = 1 = y', z' = 0$, we can deduce that the inputs were $x = y = z = 1$.

Actually the Petri gate is even *self inverse*: if we connect two instances, then the original input is reestablished. To see this it is of cause sufficient to follow the signal on the undermost z-line. At the output of the second gate we get $(z \oplus xy) \oplus xy = z \oplus (xy \oplus xy) = z + 0 = z$.

Information Flow Graphs An information flow graph consists of an arbitrary combination of Petri gates and additional negation elements, here represented graphically by a small circle. A negation-element $v -\!\!\bigcirc\!\!- \bar{v}$ transforms the input signal v into the *complementary* output signal $\bar{v}$, i.e., $\bar{0} = 1$ and $\bar{1} = 0$. Figure 5.5 shows an information flow graph built according to these rules.

Fig. 5.4 The Petri gate

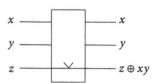

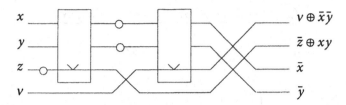

Fig. 5.5 Information flow graph. The *horizontal line* above the variables denotes negation

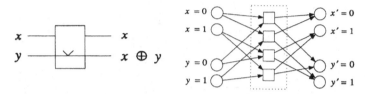

Fig. 5.6 Variant of the Petri gate, called "Pfeilfunktion P_1" by Petri. *To the right*: Unfolding into Petri net

Every information flow graph represents a reversible transformation of input-into output-values. As Petri has shown, also the converse is true. Information flow graphs are in fact sufficient to perform arbitrary Boolean computations of this type. We have the following

Fundamental theorem of reversible logic (Petri 1965 [7]). *Every Boolean bijection of an arbitrary arity n can be realized by an information flow graph with $n + 1$ inputs. The additional signal line is used only for intermediate computations. At the output it has reassumed its initial value.*

Information flow graphs can be directly unfolded into Petri nets. This is illustrated in Fig. 5.6 for a simplified example gate.

The Petri gate is not the only possible basic building block. In *Interpretations of Net Theory* [10], Petri for instance chooses another gate that he calls *Quine transfer* (after the philosopher and logician Willard Van Orman Quine). The Quine transfer is a conditional exchange of the form $(x, 0, z) \mapsto (x, 0, z)$, $(x, 1, z) \mapsto (z, 1, x)$. The values of x and z are exchanged if and only if $y = 1$. The value of y is not influenced by the gate.

The Quine transfer distinguishes itself by a particularly aesthetic unfolding into a Petri net. Figure 5.7 shows a representation, hand-drawn by Petri.

Reversible Switching Circuits Closing of the signal lines to cycles, permits delay elements to be represented by information flow graphs. By inverting the signal flow, it is possible to construct switching circuits. In the article *State-Transition Structures in Physics and in Computation* [18], Petri briefly hints at the possibilities of such

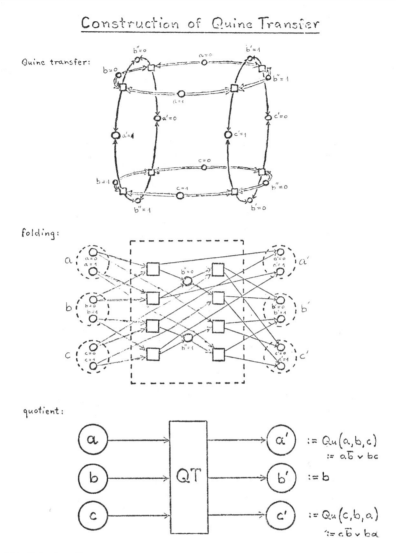

Fig. 5.7 Quine transfer

extended information flow graphs: "Several blueprints for fully-fledged computers exist in the graphical level-3 [flow graphs] notation."

In such statements, a peculiarity of Petri's concept of the world becomes apparent. As this author can confirm from numerous personal conversations, in his head an extensive range of details for the construction did in fact exist; but for him the essential part of the work was then already accomplished. Much of that what he had in mind, he unfortunately never put down in writing.

Fig. 5.8 At the Petris' home in 1978. *From left to right*: Petri, Hartmann and Helga Genrich, Joyce Fredkin

Impact on Theory ... The basic idea of information flow graphs was taken up by leading researchers worldwide. The Petri gate became known in the USA under the name *Fredkin gate*, the Quine transfer as *Toffoli gate*. In fact, from these approaches a whole field of research arose, called *conservative logic*.

A series of *International Workshops on Reversible Computing* has been arranged regularly since 2005. In the announcement to the first workshop, a lecture to be given by the "reversible logic pioneer" Ed Fredkin was highly recommended. Now, the background to Fredkin's pioneer activity can undoubtedly be traced back to his discussions with Petri, whom he visited in 1978. Figure 5.8 shows (from left to right) Petri, Petri's close co-workers Hartmann and Helga Genrich and Fredkin's wife Joyce at the Petris' house.

... and Practice The idea to embed all computations reversibly did not always meet with unqualified approval from practitioners. Petri's son Tobias recalls: About 1985 the leading engineers in a large electric and electronic company invited Petri to provide them with first hand knowledge on net theory and reliability in communications engineering. The idea of reversible logic did not immediately convince them: Why compute backwards—the past is surely well known? They were of the opinion that signals that had already been processed according to plan, were now obsolete and no longer of any value.

Before them on the table lay a large drawing of a net; the engineers used coins as tokens, pushed them along the net to the confluence of two paths, where they then only let one coin continue beyond the node. Now Petri declared: "If you say that this signal is done with and has lost its value, then I can take it away."—and perkily put the coin, which did not belong to him, in his own pocket.

Fig. 5.9 Astonishment over
a found penny?

The Come-From Command Also around 1985, in an article in the April issue
of the *Communications of the ACM*, a professional magazine published by the
American Association for Computing Machinery, the authors addressed the at the
time highly controversial issue of the *go-to* command in programming languages.
They proposed to enhance the comprehensibility of jump instructions by the
introduction of additional *come-from* commands. Petri commented on this idea with
a chuckle: "They obviously intend this as a particularly hilarious April fool hoax."

GMD: Home of the Petri Nets

6

Contents

On April 23, 1968, the "Gesellschaft für Mathematik und Datenverarbeitung" (GMD, Society for Mathematics and Information Technology) was founded within the grounds of the Birlinghoven castle in Sankt Augustin near Bonn, at that time the capital of Western Germany.

The castle had been built around the year 1900 by a Rautenstrauch family of entrepreneurs, and was soon after sold to the banker Louis Hagen from Cologne. In 1959 the Deutsche Shell AG set up an institute for basic research in chemistry on the premises. In 1968 the castle was acquired by the West German State, the Federal Republic of Germany.

The development of the GMD is documented in detail in the volume *Informatik und Großforschung. Geschichte der Gesellschaft für Mathematik und Datenverarbeitung* [25] (Computer Science and Large Scale Research. History of the GMD). In the present chapter we can therefore limit ourselves to a brief overview.

© Springer-Verlag Berlin Heidelberg 2015

E. Smith, *Carl Adam Petri*, DOI 10.1007/978-3-662-48093-9_6

Fig. 6.1 Birlinghoven Castle. Until the end of the renovation works in the institute buildings, Petri's office was located temporarily in the white tower annex on the upper left, which of course occasionally gave rise to ironic remarks

6.1   Bonn or Munich?

The objective of the foundation of the GMD was to carry over the concept of large scale research, which had already proved to be successful in the field of nuclear energy, to data processing. Two possible locations were discussed, Munich in Bavaria and Bonn. A vigorous advocate for Munich was Friedrich Ludwig Bauer, a renowned professor at the Technical University of Munich. The final decision fell in favor of Bonn, however. The argument was based on two reasons: The first was that the Ministry for Research, located in Bonn, wanted the GMD nearby. The Deutsche Forschungsgemeinschaft (DFG, German Research Foundation), also located in Bonn, was like-minded. The second was that in the decision-making meetings in the DFG, Petri could convincingly present his concept of a future-oriented computer science, in which he explicitly foresaw computers as *media of communication*. This insight clearly benefited from his experiences with the ARPA-network.

6.2   Bonn

The result was that the "Institut für Instrumentelle Mathematik" (IIM) at the University of Bonn was enlarged and converted to the GMD. The initial managing directors of the GMD were Petri's academic supervisor Heinz Unger and Ernst Peschl, the latter also mathematics professor at Bonn University. Petri was appointed head of the "Institut für Informationssystemforschung" (IFS, Institute for Information System Research), later renamed "Institut für Methodische Grundlagen der Informationstechnik" (IMG, Institute for Methodic Foundations in Information

Technology), in the first instance for 3 years. 1971 the appointment was renewed for another 5 years.

In 1973 Petri was offered a computer-science chair at the University of Dortmund. He declined, but used the offer to renegotiate his position in the GMD. In this way he managed to obtain better conditions for the institute than had been possible at a university. His appointment as head of the institute was extended, he received the status of a civil servant for life. He remained in office until he retired in 1991, all in all for 23 years.

6.3   The GMD Under Criticism

Even early in its existence, the GMD was subject to substantial criticism. In 1972 the "Bundesrechnungshof" (Federal Audit Office) objected that the GMD was almost exclusively oriented towards fundamental mathematical research, instead of performing the originally assigned task, which consisted of supplying the federal administrations with computer programs for the collection and processing of data. Moreover, it was claimed that the GMD was not using its substantial financial resources—provided by federal funds—in an economically responsible manner.

Occasional rumor had it that these attacks were launched by interested parties, who perhaps would have preferred to see the first large scale research on data processing in some city other than Bonn.

A similar attack was initiated in the mid 80s, when all of a sudden various press media simultaneously and unanimously criticized the "overpaid GMD employees neglecting their assigned duties", with the conclusion that the public funds granted to the GMD could have been put to much better use elsewhere.

In fact, as early as 1969, plans were discussed within the Ministry of Research and Technology, to reorient the GMD towards engineering science and software technology. In 1970 this opinion did win through, and Prof. Unger was removed from office.

With regard to "uneconomical" use of funds, the complaint had been made that, amongst other things, the computer terminals were not used to their full capacity. In a comment Petri had replied that in his opinion terminals were not to be considered as production equipment in the usual sense, for which the notion of maximal work load applies, but rather as *communication media*. In an ironic tone he pointed to the fact that likewise the telephone sets were not used to full capacity, and—to his knowledge—the fire extinguishers had never been used at all.

Petri's institute survived the storm unscathed. The 70s were to become the blooming period of net development.

6.4   Condition-Event Systems

In the early 1970s, the net class most important to Petri, the so-called *condition-event systems*, was brought to its final form. The term Petri net is used to encompass the whole class of net models, but also in particular to denote those nets where places may hold multiplicities of tokens. In contrast, a *condition* is either satisfied or not. In graphical representation, thus only *one* token may appear on each place. We shall return to nets with multiplicities of tokens in Sect. 6.9 below.

The unwieldy impersonal term condition-event system has unfortunately prevailed, even though other names have repeatedly been suggested, such as *elementary Petri system*. A compromise that has come into widespread use is the abbreviation CE-system.

The central ideas of CE-systems were presented to the scientific public in September 1973, at a symposium in the High Tatras in Czechoslovakia [9], and then published in an internal report of the GMD, *Interpretations of Net Theory* [10], 1975.

Following the Austrian philosopher Ludwig Wittgenstein, Petri explains the foundations of condition-event systems as follows: "A case is a collection of *conditions*. ... A process is a collection of occurrences of events. ... A process changes that which is the case." And crucially: "We permit concurrency of event occurrences."

A CE-system basically consists of an underlying net (for illustration see for instance the examples in Chap. 1, disregarding the tokens). A *case* represents the current state of the system, depicted graphically by a corresponding token distribution.

Example

Figure 1.1 in the Introduction shows a system with two distinct cases, one case consisting of the condition "off", the other one consisting of the condition "on". Figure 1.9 shows a CE-system in a case consisting of the conditions a_{id}, c_a and b_{id}, in set notation $\{a_{id}, c_a, b_{id}\}$.

The dynamics in a CE-system now follow the observation, that cases are transformed into new cases according to the rules of the token game.

Example

In Fig. 1.9 the current case could be transformed, say, into the case $\{a?, c_a, b_{id}\}$ by the occurrence of w_a, or into $\{a_{id}, c_b, b_{id}\}$ by the occurrence of l_b.

In general, a CE-system consists of a *net* and a *class of cases* satisfying certain requirements, the most basic one being that the case class consists precisely of those token distributions that can be reached by successive occurrences of events, beginning with a distinguished initial case.

As already discussed, for Petri the concept of time reversibility on a small scale is of particular importance. In a stricter sense he therefore demands that the case class must also be closed under backward occurrences of events, more precisely, even under all mixed sequences of forward and backward occurrences.

The most natural and adequate description of dynamic *behavior* in CE-systems is given by the *non-sequential processes*, already mentioned in the Introduction.

Example

For instance, the process in Fig. 1.10 transforms the case $\{a_{id}, c_a, b_{id}\}$ into $\{a_{id}, c_b, b!\}$.

For a deeper understanding of distributed systems, it is important to note that processes correctly mirror *both* dependence *and* independence between events.

6.5 Confusion

As already emphasized in Chap. 1, the three basic notions *concurrency, sequence* and *conflict* suffice to describe the behavior of net systems.

Through a peculiar interplay between all three of them, a further remarkable constellation may emerge, introduced by Petri under the name of *confusion*. Figure 6.2 illustrates the typical interdependence-structure pertaining to such a confusion.

In the case $\{a, c\}$ represented here, both of the events A and C are enabled. Through occurrence of C we may arrive at the case $\{a, e\}$. Since the condition b does not hold, no conflict-resolution was required between B and C. By occurrence of A the system can now change into the case $\{b, e\}$.

Returning once more to the initial case $\{a, c\}$, we now assume that A occurs, and the system thus changes into the case $\{b, c\}$. Now, as before, C may occur and transform the system into the same final case $\{b, e\}$, but with a significant difference: In order to reach this constellation, a *conflict* between B and C had to be resolved.

We may thus arrive at the same result, either *without* or *with* conflict. What characterizes a confusion, is that it is afterwards no longer possible to determine whether a conflict resolution was required or not.

Formally, a confusion arises in an interplay between three events A, B, C, for which (1) A is a causal predecessor of B, (2) B is in a *structurally potential* conflict with C. (3) Whether the conflict becomes actual, is determined by an event A concurrent with C.

As we shall discuss below, confusion turns out to be one of the most interesting phenomena in distributed systems, in this author's opinion arguably *the* most fundamental one of all. When Petri introduced the notion of confusion, the net language was the first system model in which it could be at all precisely identified and formalized, and even today it is probably one of net theory's unique selling points.

Unfortunately, even in the net literature, confusion still appears to have only a shadowy existence.

Fig. 6.2 Confusion

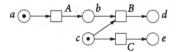

Early Views on Confusion Confusion was given so little attention probably because Petri himself initially believed that it could be avoided. For a long time he saw confusion only as an indicator of badly designed systems. In an article *Concepts of Net Theory* [9] from 1973 he remarks: "Fortunately it is easy to avoid confusion by applying well known construction principles." In *General Net Theory* [13] from 1977, he expounds in more detail: "When we encounter such a situation we may conclude that we have drawn the boundary between the system and its environment in an awkward manner, and that we should draw it somewhere else, in order to reduce confusion to mere conflict resolution. It follows that we really don't need to set up a theory of confusion—which would indeed be difficult to do. It has, in fact, been tried more than once and seems almost impossible."

The deeper reason for Petri's opinion appears to be based on his view—discussed in Sect. 5.5—that conflicts do not exist in a comprehensive net, but only appear within a system, when the border between the system and its environment has been drawn in a way that the condition necessary to decide the conflict, has come to lie outside of the (partial) system.

With respect to confusion, then the obvious question arises, how can a border between a system and its environment exist, such that (1) on the one hand, the condition required for conflict resolution is located *outside* of the system, but (2) on the other hand, there is possibly *no such condition at all*, simply because there is no conflict. To resolve this contradiction, it is clear that significantly more elaborate assumptions have to made on the relationship between system and environment.

Confusion in the Literature For his reluctant attitude towards confusion, Petri received whole-hearted support from his colleague Anatol Holt. In an article *Concurrency and Choice* from 1977, Holt recognizes that in a confusion there is an "extra factor" contributing to conflict resolution, and contends: "Thus an extra causal factor, not represented explicitly in choices ... has entered the scene, *namely the factor of relative timing. ...* This 'extra factor' is both technically and philosophically unacceptable. It is technically unacceptable because it destroys the possibility of tracing the outcomes of choices to the outcome of other choices ... – one of the key objectives, in my opinion, of an adequate system model."

Regarding the philosophical significance of the time-factor, Holt remarks: "Communication and only communication establishes causal connections between choices. Concurrency was to express the relative freedoms that remain in the light of these relative causal constraints." According to Holt, something that may influence causality must be an element of communication, and that is not the case for the time factor.

In essence, Holt's conclusion amounts to the view that what *should not* exist, actually also *does not* exist. This completely accords with a well-known observation by the German satirical poet Christian Morgenstern in a poem "Die unmögliche Tatsache" (The impossible fact): "Weil, so schließt er messerscharf, nicht sein kann, was nicht sein darf." (For, he reasons pointedly, That which must not, can not be.)

In Petri's closer circle of colleagues, the belief that confusion was to be—and could be—avoided, often gave reason to concentrate research activities on special limited classes of nets, where confusion was in fact excluded by construction.

Confusion is Unavoidable In the meantime it turned out, however, that in every real system, confusion is actually *unavoidable*, and appears wherever different *partial systems are synchronized.* After this had been demonstrated mathematically by this author [24], Petri, in a private communication with him, commented: "Before, I used to believe that confusion can be avoided in modeling. Now I claim the opposite."

In Chap. 4 we mentioned that a subtle error might have crept in into the apparently conflict-free construction of the OR gate in Petri's thesis. Now we can provide the reason for that remark: In light of the above result, there must exist a confusion in the construction, hence *a fortiori* also a conflict.

In a certain sense, the situation can almost be considered to be tragic (in the classical meaning of the word). The tool that like no others permits the precise distinction between causality and independence, exactly for this same reason, permits the characterization of situations where causality and independence *cannot* be distinguished clearly!

One may perhaps be reminded of the ancient Greek philosopher Pythagoras, whose philosophical foundations were put in doubt by one of his own results. Pythagorean philosophy is based on the assumption that all magnitudes in nature occur in definite proportions to one another, corresponding to ratios of whole numbers. The famous Pythagorean theorem, however, has exactly the opposite consequence: In a square where the edges are of length 1, there is no way to express the length of the diagonal by a ratio of whole numbers.

To resume the discussion, one could say that the cosmological foundations of Petri's philosophy, on which the nets are founded, cannot be discarded lightheartedly. In practical modeling they should however not always be adopted categorically.

In the following we illustrate confusion in some typical contexts.

Fig. 6.3 The philosopher Pythagoras between theorem and collapsed pillar. Drawing by Petri around 2002

Dijkstra and the Philosophers There is no doubt, confusion is a rather unpleasant phenomenon. But often we only have a choice between cholera and the plague. As Petri recalled years afterwards, around 1962, he, the famous Dutch computer scientist Edsger Dijkstra, and a third colleague by the name of Meisel, were sitting around a round table on a hot summer afternoon. A bowl of fruit salad had just been served. It so happened that all three of them reached for the serving spoon contemporaneously. The resulting triple-conflict could of course be resolved by common sense, but it also led to a discussion of the notion of *glitch*, a short spurious pulse within a system, which however disappears again by itself, and thus makes troubleshooting difficult. In electronics, the term usually refers to short-lived fallacious values in logical circuits.

Dijkstra and Meisel were of the opinion that glitches were unavoidable by the laws of nature, the best one could do was to minimize the consequences. Petri substantially agreed, but said that he still had to think about it some more. He came up with the solution 23 years later: The problem was confusion, not conflict.

Dijkstra returned home and within about a week generalized the problem, and came up with the so-called *Dining Philosophers Problem.*

Five persons are sitting at a round table to eat spaghetti. Between any two of them, there is a fork. For eating they need *two* forks, both the left and the right. In the worst case, each philosopher reaches for his right fork first. The system thus runs into a *deadlock*, nobody can eat.

For each participant the action cycle is as illustrated in Fig. 6.4: If both the right and left forks are available (conditions *r* and *l*), he can pick them up and begin eating (event *b*). He terminates the eating cycle by putting down the forks (event *z*). The additional arrows leaving *r* and *l* denote that corresponding situations exist for the neighbors, and that they thus are in conflict with each other. However these are *pure* conflicts, not embedded in any confusion.

The possibility of deadlock obviously results from the fact that the forks are picked up as *two independent* actions. Deadlock can be avoided, if pick-up of both forks can be conceived as belonging to *one atomic action.* However this only replaces one evil with another. Instead of deadlock, the system will then inevitably contain the possibility of confusion. This is illustrated in the system segment shown in Fig. 6.5.

The philosopher to the extreme left—let us call him Confusius—intends to start eating, both forks are available (conditions g_1, g_2). First, however, he still wishes to conclude a thought he has in mind. He believes there is no hurry, because his neighbor cannot take up the shared fork 2, since fork 3 is in use. When he then finally decides to pick up the forks, it is too late. In the meantime the neighbor next

Fig. 6.4 Segment from the philosophers system with independent pick-up of forks

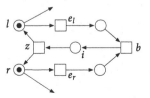

Fig. 6.5 Segment from the philosophers system with coincidental pick-up of forks

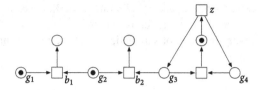

but one to the right has put down his forks, resulting in a conflict; his immediate neighbor decides to take advantage of this. In net terms this corresponds exactly to the fact that the three events b_1, b_2, z form a confusion in the case depicted.

In technical real-world situations—for instance in operating systems—standard algorithms often use so-called *semaphores* in order to avoid deadlocks. In all such constructions then, unavoidably, lurks the danger of confusion.

In situations where one knows from experience that there is little danger of deadlock, occasionally a very simple method is used. According to Andrew Tanenbaum in his standard reference on operating systems, then the *Ostrich algorithm* is often considered sufficient, which consists in ignoring the problem, and pulling the plug, should it nonetheless happen.

Confusion in Computer Science Applications ... In practice, confusion is well-nigh ubiquitous. Returning to the model of the printer access in Fig. 1.9 in Chap. 1, it is easy to see that confusion cannot be excluded in the direct vicinity of the condition c_a. It suffices to project Fig. 6.2 as follows: A to w_a, B to h_a and C to l_a. The reader is invited to verify that a similar confusion involving the condition c_b is also reachable.

Given that we have to accept confusion, we also have to consider how to cope with it. A system like Fig. 1.9 will always contain confuse situations, regardless of how technically complex the realization may be. A possible (and arguably also the only) approach to live with it, could consist in assigning appropriate *priorities* to conflicting transitions in the immediate vicinity of the confusion. More about this in Sect. 10.9.

...and in Everyday Practice But also in real-life situations, hard to find errors can often be explained by tracing them back to a confusion. Within a joint European research project on office organization, a group of Petri's co-workers, led by Gernot Richter in the mid-1980s encountered the following problem:

In a public kindergarten in England the rule was, that a child had to vacate its place if the parents fell behind with the fees. The head kindergarten teacher however could at her discretion in case of hardship refrain from expelling children, even when the parents were in default. In the annual account the supervisory board reprimanded her for having used the exceptional rule much too often. The teacher objected vehemently. In the following analysis of the procedures, it turned out that the background for the different assessments comprised a confusion, which corresponded exactly to Fig. 6.2, when the net is interpreted as follows:

C signifies that the kindergarten teacher accepts the child for the next month according to the *general rule*. The event A represents the arrival of the missing-payment notice. It turned out that the central city treasury was not aware of the exception rule, and usually only sent the missing-payment notice to the teacher after 3 months (occurrence of event *A*). Hence the precondition for the *exception rule* (event *B*) was not satisfied, so that the teacher had neither reason nor possibility to apply it.

6.6   Contact

Events can occur when they are enabled: the preconditions must hold, *and* additionally no postcondition may already be satisfied. For instance, the event *A* in Fig. 6.6 is *not* enabled, although both preconditions *a* and *b* are satisfied, because also a postcondition *c* is already holding.

This particular form of the enabling-rule stems from Petri's insight in his thesis, that an action *transforms* input- into output-signals. A condition that is already *satisfied*, cannot be *made* to hold by an event. Petri calls this the *extensionality principle* in nets, meaning that an event is already completely determined by the changes it effects. If *A* in Fig. 6.6 were to occur, it would transform the case $\{a, b, c\}$ into $\{c, d\}$. Arguing backwards, by the extensionality principle we could deduce that in reality an event had occurred that transforms $\{a, b\}$ into $\{d\}$.

In the article *Concepts of Net Theory* [9], Petri denotes the situation illustrated in Fig. 6.6 as *contact*, and explains it like this: "An elementary particle can not occupy the space already occupied by another one, but a car unfortunately can, and so can a record of data."

What Petri means is that a contact situation is physically impossible on the lowest level, whereas on a higher level it indicates a possible malfunction. He notes that for an event in a contact situation the following holds: "We say that an occurrence of *e* would violate our definitions. ... On a higher level a violation means a violation of the rule of the game." Since he wants his model to be applicable at all levels of description, he cannot really feel sympathetic to the concept of contact. As in the case of confusion, he advocates the view that contacts are a consequence of badly designed models.

Petri's understanding of contacts has often been interpreted as a demand to exclude them from modeling altogether, which is in fact mathematically easily possible, by the introduction of so-called *complement*-conditions. In that way a behaviorally equivalent system is generated, where now however each *post*-condition responsible for a contact is supplemented by an additional *pre*-condition $\bar{a}$ that does *not* hold if and only if *a* does.

A collateral damage of place-complements is however, that the expressive power of the net model is unnecessarily—but significantly—reduced.

Fig. 6.6 Contact

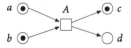

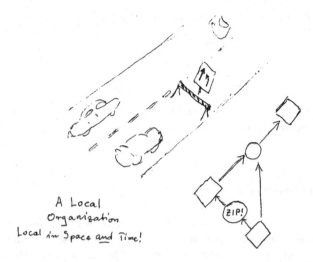

Fig. 6.7 Danger of contact. Drawing by Petri around 2002

Fig. 6.8 Railway system with trains in track sections *a* and *c*

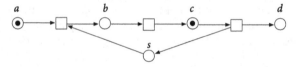

Fig. 6.9 Railway system in Fig. 6.8 with collision protection

---

Example

To illustrate, we consider the model of a railway system shown in Fig. 6.8.

The track is divided into sections *a* to *d*. At present there are two trains in sections *a* and *c*. If the posterior train now enters section *b*, there is a possible danger of collision with the train in the neighboring section *c*. The situation is represented in the net by the contact in the case {*b*, *c*}. The potential danger *noted* in the model can then be taken as a warning signal to implement *technical* measures in the *real* system, in order to prevent an actual collision.

The model of an appropriately modified system, augmented by a collision protection device, could then look like in Fig. 6.9. The condition *s* indicates whether *both* sections *b* and *c* are clear; a train is only allowed to drive into section *b*, if there is no train in *c*.

As illustrated in the example, it does not appear sensible to rashly *define away* contact situations from the net model. On the contrary, they are of great use in the

Fig. 6.10 Synchronic
distance $\sigma(a, b) = 2$

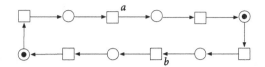

detection of potential critical behavior. Physically, two objects *cannot* be at the same place at the same time. It may however be of crucial importance to prevent even the *attempt*—such as the intrusion of a train into an already occupied track section.

6.7   Synchronic Distance

When two events may occur concurrently, it is natural to ask for a *quantitative* measure for their relative independence or coupling. To this end, Petri introduced the notion of *synchronic distance* in *Interpretations of Net Theory* [10]. The synchronic distance $\sigma(a, b)$ between events a and b states how often one the two events may at most occur before it is the other one's turn.

In the system in Fig. 6.10, taken from *Interpretations* [10], we get for instance $\sigma(a, b) = 2$. (More generally Petri defines the synchronic distance between *sets* of events.) The idea was—in mathematical terms—to derive a discrete metric (i.e., with whole numbers as values) from the synchronic distance. The approach did not, however, live up to its initial hopes, mainly because the triangle inequality $\sigma(a, c) \leq \sigma(a, b) + \sigma(b, c)$ in general only holds for comparatively simple systems. Consequently the idea was not followed up with any substantial effort.

6.8   Enlogic Structure

In *Interpretations of Net Theory* [10] and also in *Non-Sequential Processes* [12] we find a significantly more sustainable analysis- and specification-method. It relies on the introduction of an additional type of transitional form—*hypothetical virtual transitions*, and the investigation into what implications they would have on system behavior.

Example

To illustrate, we return to the printer system in Fig. 1.9 in Chap. 1, and imagine an additional transition h'_a similar to the real event h_a, except that h'_a has c_b as an additional output condition. If now h'_a could occur in the case $\{a?, c_a, b_{\mathrm{id}}\}$, then it would cause the condition $a!$ to be satisfied, the exclusive access to the printer would be granted to agent a. But the condition c_b would then also hold, and thus fallaciously indicate accessibility to agent b.

Petri calls such a transitional form, which could transform *at least one* case into a non-case (corresponding to a token distribution that is not actually reachable), a *violation*. In contrast, a transitional form that would *always* transform cases into cases, he names *process extension*. All in all he distinguishes 16 types of such transitional forms. They determine the so-called *enlogic structure* of the system. Petri coined this term in analogy to the notion of *entropy* used as a measure of disorder in thermodynamics.

Facts Of particular interest are the so-called *facts*, "dead" transitional forms, which could not occur in *any* case of the system. They correspond to properties that remain true invariant to state changes; in Petri's words: *factually true*.

To illustrate, imagine Fig. 6.9 above, augmented by a transition with preconditions b and c. It could never occur, since by construction b and c are never simultaneously satisfied.

Facts may also be of use in the *specification* of *desired* behavior. To distinguish, these are often also called "to-be-facts". For instance, if we were to add a transitional form as above with input conditions b and c to Fig. 6.8, it would be a violation, but it could also be understood as a to-be-fact, with the meaning: Please *modify* the system, so that it actually becomes an (is-)fact.

Facts describe system invariants with respect to propositional logic. As a subclass, the facts include the so-called impure transitions, where input- and output-conditions are not disjoint. These correspond to the tautologies of propositional logic, with the system conditions as propositional variables.

Facts can be manipulated using the standard derivation rules of propositional logic, from which, in turn, a comprehensive calculus of invariants in nets can be derived. This was worked out in detail in the 1970s at the Petri-institute at the GMD, chiefly by Gerda Thieler-Mevissen.

6.9   Place-Transition Nets

In its general sense, the term Petri net is used collectively for all net models that can be traced back to Petri's basic idea. The essential features usually associated with Petri nets are: The underlying structure consists of a bipartite directed graph. Elements of the one sort S are interpreted as carriers of data, elements of the other sort T as actions, by which the data on the connected S-elements are processed.

The notion Petri net is, however, also used in a more specific sense, for a net class that Petri originally introduced in *Interpretations of Net Theory* [10] under the name *place-transition nets* (PT-nets). In contrast to condition-event systems, in PT-nets the places can be marked with a number > 1 of tokens. Correspondingly, a transition may withdraw or add more than one token in a single action. Note also the different manners of speaking: In CE-systems 'events occur', in PT-nets it is more customary to talk about the *firing* of *transitions*. Also the term *case* is not used, rather we speak of a *marking*.

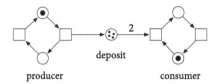

Fig. 6.11 Producer-consumer system

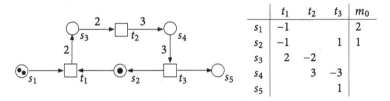

Fig. 6.12 PT-net, incidence matrix A, and initial marking m_0

Example

Figure 6.11 shows a producer-consumer system, in which the producer stores his products in a deposit, from where the consumer can then withdraw them.

At present, the deposit contains four product units, represented by the four tokens. The consumer may withdraw products in sets of two units, indicated by the inscription 2 on the connecting edge.

Incidence Matrix PT-nets are often represented by their so-called *incidence matrix*.

Example

Figure 6.12 illustrates the idea with a simple example.

The incidence matrix A consists of entries that describe the effect of the transitions. A firing of t_1, for example, withdraws one token from each of the places s_1 and s_2, and instead adds two tokens to the place s_3. To the extreme right the initial marking m_0, corresponding to the token distribution shown, is written as a column vector.

S-Invariants The incidence matrix is used in one of the main tools in net theory, system analysis with S-invariants, also known as the workhorse of net theory. An S-invariant is a column vector i with an integer entry ≥ 0 for each place, such that $i^T \cdot A = 0$ with normal matrix multiplication.

Example

In Fig. 6.12, we have for instance the S-invariant $i^T = (0, 6, 3, 2, 0)$.

S-invariants are particularly useful in the verification of *non*-reachability. The reason is that for any *S*-invariant and any marking m reachable from the initial marking m_0, the scalar product $i^T \cdot m_0$ is always equal to $i^T \cdot m$.

Example

In the example above, the marking $m^T = (2, 1, 0, 1, 0)$ is *not* reachable, because

$$i^T \cdot m_0 = (0, 6, 3, 2, 0) \cdot \begin{pmatrix} 2 \\ 1 \\ 0 \\ 0 \\ 0 \end{pmatrix} = 6 \neq 8 = (0, 6, 3, 2, 0) \cdot \begin{pmatrix} 2 \\ 1 \\ 0 \\ 1 \\ 0 \end{pmatrix} = i^T \cdot m.$$

The reason for the term "invariant" is this: An *S*-invariant assigns whole-number weights to places, such that the *weighted sum remains invariant* under transition firings. In fact, for non-reachability analysis it is not even necessary to require the vector i to consist merely of whole-number entries ≥ 0. Actually, for *any* solution i of the equation system $i^T \cdot A = 0$ we get $i^T \cdot m_0 = i^T \cdot m$ for every marking m reachable from the initial marking m_0.

"Petri Nets" Petri himself is only marginally interested in PT-nets. His own most important use appears to be in the characterization of synchronic distances between events a and b in condition-event systems, for instance in *Interpretations of Net Theory* [10] and *General Net Theory* [13]; to determine synchronic distance, we can insert an additional place between a and b and ask for the minimum number of tokens it must hold, in order not to impair the behavior of the original system.

In a certain sense, PT-nets are even incompatible with Petri's fundamental postulates. Already in his thesis he emphasizes, that no real memory element can accommodate arbitrarily large natural numbers. But with places of unlimited capacity, the incompatibility to realistic physical limitations sneaks in through the backdoor.

In an unpublished note from around 2002, he formulated his discomfort with place-transition nets as follows: "We avoid 'place' as used in net literature. 'Places' are supposed to be capable of infinite size; and thus admissible as stores for natural numbers. This has led to grave misconceptions and errors in the assessment of the theoretical potential of net theory, and has severely harmed its development. The term evokes misleading associations, and is really a mistranslation of the German 'Stelle', which has been used to suggest *ST*ate *EL*ement by abbreviation. 'Place' has no place in net theory."

6.10   Dissemination

As mentioned, net theory and applications were initially strongly promoted in the early 1970s in the USA, especially by the Computation Structure Group at MIT. It was also at the MIT, that the first conference on nets was arranged in 1975.

Advanced Course Hamburg in 1979   In October 1979, the breakthrough reached Europe, when a 2 week *Advanced Course on General Net Theory of Processes and Systems* was held in Hamburg, Germany. In the course, the state of the development was presented to the general scientific public. Prof. Wilfried Brauer was the initiator and manager of the course, and he also subsequently remained one of the major advocates of Petri nets. Brauer had been active at the University of Bonn in the 60s, where, after his doctorate in mathematics, he dedicated himself to automata theory. He thus already had a long familiarity with Petri nets.

More than 100 participants from 17, mainly European, countries were present at the course. The contributions appeared as volume 84 of the series *Lecture Notes in Computer Science*, published by Springer-Verlag. (At the time of writing the series is coming up to 9000 volumes.)

The leading researchers in Petri's GMD-institute at that time, Hartman Genrich, Kurt Lautenbach and P. S. Thiagarajan (in full P. S. stands for Pazhamaneri Subramaniam) documented the foundations of net theory in a 130 page article *Elements of General Net Theory*. Genrich and Lautenbach had already worked with Petri in the Institute for Instrumental Mathematics at the University of Bonn, Thiagarajan came to the GMD from the Massachusetts Institute of Technology (MIT) in 1975.

Petri gave two lectures, *Introduction to General Net Theory* and *Concurrency*. We shall return to the topics treated there in more detail in later chapters.

The German computer-pioneer Konrad Zuse emphasized the relevance of Petri nets from an engineering point of view. The relationship between Zuse and Petri will be discussed in Sect. 7.3.

Repercussions   In the aftermath of the Advanced Course, what was hinted at above grew in prominence: Practitioners and theoreticians alike eagerly adopted Petri nets in the form of place-transition nets. They were now given the means to represent systems with distributed resources in an intuitively attractive manner. At the same time, mathematical methods started to appear, which could deal with the already mentioned reachability analysis, among others. The matrix representation enabled the application of linear algebra theory to systems analysis. The development of powerful net-based simulation tools had its beginning.

The wave was so forceful, that the condition-event systems of "doctrinal theory" were pushed into the background, and were rather considered as a special case of PT-nets, comparable to the view in modern programming languages, where logical values are represented by the numbers 0 and 1, and thus, so to speak, understood as merely a special case of general arithmetic.

The fundamental level in nets, based on the logic of conditions "holding", "not holding" was mainly worked on in the closer environment of the Petri-institute. Condition-event systems in this sense deal with *alterable truth-values* in logical propositions. This approach was then extended to nets where the places became *predicates with alterable extension*, corresponding to the step from propositional to predicate logic. The main contributors were Petri's close co-workers Hartmann Genrich and Kurt Lautenbach. Though logically sound, for practitioners there still remained the view that the bivalence "yes-no" was too limited. What caught on, were nets similar to PT-nets, with multiplicities of tokens, which however could, in addition, be of different *types*, descriptively characterized by different "colors".

The various practically oriented net classes are documented extensively in the literature, so we can continue to concentrate here on Petri's own work.

The Annual Petri Net Conference In 1980 the *First European Workshop on Application and Theory of Petri Nets* was organized in Strasbourg, France. Since then the conference has taken place annually. The second, in 1981 in Bad Honnef, was so to speak a home game in the immediate vicinity of the Petri-institute in Sankt Augustin. Later the conference was renamed *International Conference on Application and Theory of Petri Nets*, to acknowledge the increasing worldwide relevance of the net approach. It was arranged outside Europe for the first time in the year 1993, in the USA, in Chicago, and since then also in Japan, Australia and China, amongst others.

Books In the early 80s the first textbooks on Petri nets appeared, in the USA *Petri Net Theory and the Modeling of Systems* by James Peterson, which was to have a large impact also outside the computer-science community. In Western Germany the first textbook was *Petrinetze. Eine Einführung* by Wolfgang Reisig, which was soon also translated into English (and later into various other languages), and became one of the standard references. In Eastern Germany, Peter Starke published his *Petrinetze: Grundlagen, Anwendungen, Theorie*.

Net Foldings, Morphisms and Topology

7

Contents

Now that net theory had proved to be ready for widespread use, it no longer required Petri's personal intervention. He could thus dedicate himself to the further development of the foundations. Ultimately, what he strived at, was a general axiomatization of concurrency and the theory of distributed systems, based on the concepts *distributedness* and *independence*. In his 1978 paper *Concurrency as a Basis of System Thinking* [16], he states that a central theme in this program is the systematic permeation of a sequence of conceptual levels and their interrelations (see Fig. 7.1).

7.1 Foldings

For the development of the program it is clearly necessary to be able to relate different net levels to one another.

In fact, we have already met a typical example for the relationship between net levels: As mentioned repeatedly, the behavior of condition-event systems may be characterized through non-sequential processes, which—again in form of a net—represent the distinct occurrences of events and conditions, as well as their structural dependencies and independencies. To recall, take another look at the system in Fig. 1.9 and the process in Fig. 1.10 in Chap. 1.

Formally, such a non-sequential process can be considered as an *unfolding* of the system. The unfolding can however also be reversed. To obtain a converse *folding* of the process in Fig. 1.10, for instance, we first have to (1) map together equally

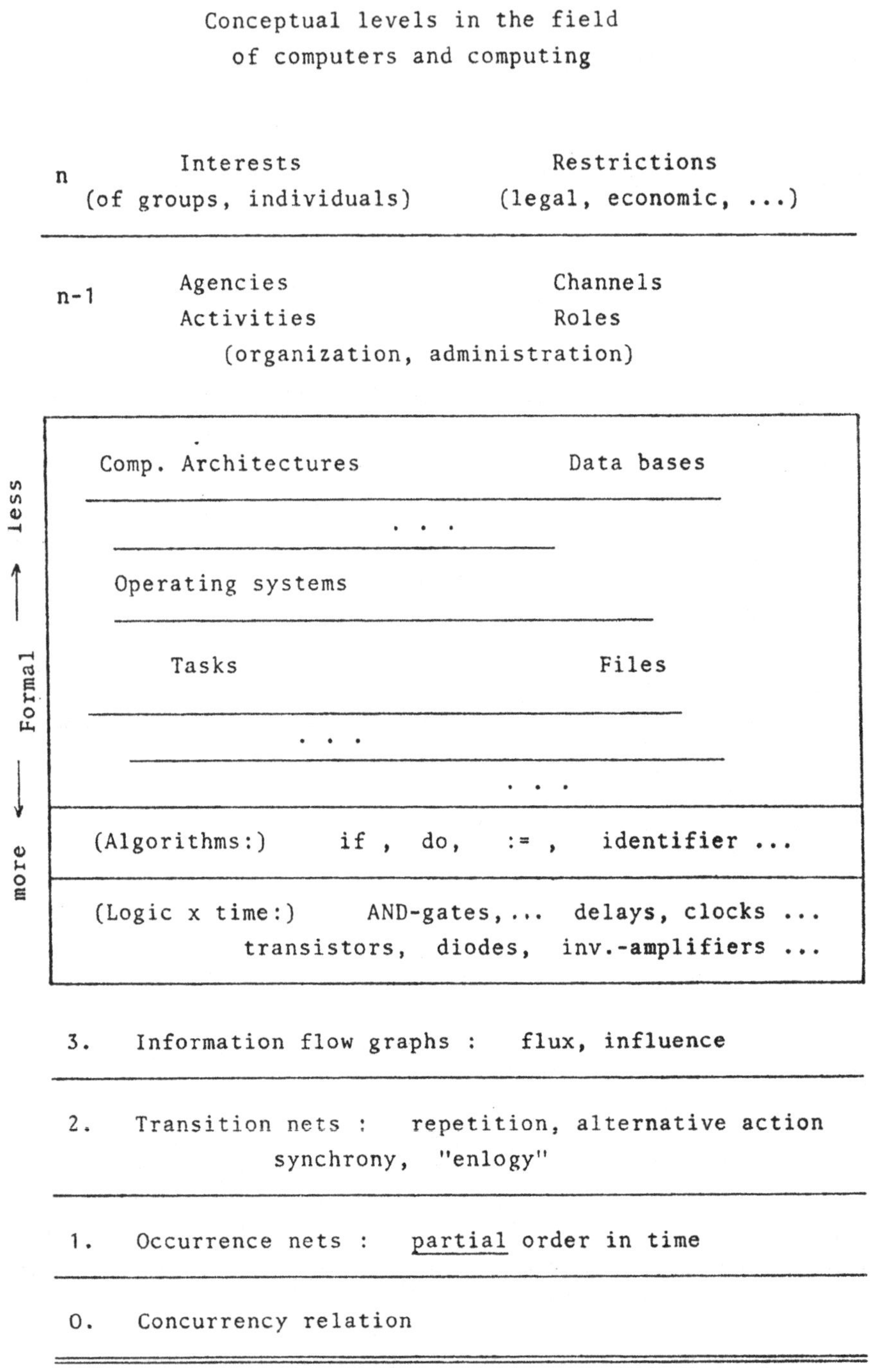

Conceptual levels in the field
of computers and computing

n   Interests   Restrictions
     (of groups, individuals)   (legal, economic, ...)

n-1 Agencies   Channels
     Activities   Roles
           (organization, administration)

less

↑

Formal

↓

more

Comp. Architectures   Data bases
                    . . .
Operating systems

     Tasks   Files
                . . .
                              . . .

(Algorithms:)   if , do,   := ,   identifier ...

(Logic x time:)   AND-gates,... delays, clocks ...
                transistors, diodes, inv.-amplifiers ...

3.   Information flow graphs :   flux, influence

2.   Transition nets :   repetition, alternative action
              synchrony, "enlogy"

1.   Occurrence nets :   partial order in time

0.   Concurrency relation

Fig. 7.1 Petri's *conceptual levels* in computer science [16]

Fig. 7.2 System obtained by folding the process in Fig. 1.10

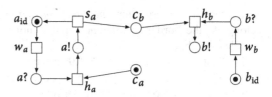

labeled elements (in this particular example only the two occurrences of a_{id} can be merged—in mathematical terms: *identified*—all the others occur only once), then (2) select an arbitrary maximal set of independent conditions within the process structure (in the example, say, the first occurrence of a_{id}, and c_a and b_{id}), and finally (3) mark these conditions in the folded net. The result will then again be a condition-event system. This is illustrated in Fig. 7.2.

The process in Fig. 1.10 does not contain occurrences of all of the original system elements, hence the folding rebuilds only a *partial subsystem* of the original one. If however the process unfolding is comprehensive enough to contain each system element at least once, then the whole original carrier system can be regained completely. The reader is invited to extend the process in Fig. 1.10 to one, which will cover the entire system.

In *Interpretations of Net Theory* [10], Petri describes the relationship between process, system and folding as follows: "The *occurrences* of conditions and events form a net …; it can be mapped … to the underlying condition event net." In *Non-Sequential Processes* [12], he speaks of "the *occurrence net* of a real process, i.e., unmarked nets where each condition has no more than one beginning and no more than one ending, such that it can be folded into the system net and thus describe a system run."

In the illustration of foldings above, we started from a given system, and showed how it could be *regained* by the folding of processes. However, we may also use processes to *specify* some desired behavior, without reference to any existing system; and then conversely to ask what a system could look like, in which the given processes could run. In many cases, it will indeed be possible to *generate* such a system from foldings.

In Sect. 6.6 we discussed *contact* situations, where an event is not enabled only because a postcondition already holds. Petri occasionally remarked that he saw the appearance of contacts mainly as a symptom of bad system design, due to improper or clumsy applications of foldings (see e.g. *Concepts of Net Theory* [9]).

7.2 Net Topology

The operations of unfolding and folding can thus be seen to be inverses of each other. In a more general mathematical sense, foldings are a special case of certain continuous mappings between nets. The idea is explained in detail for instance in *General Net Theory* [13]. Following the mathematical discipline of topology, Petri calls a subnet of a larger net *open*, if its *border* consists only of *places*. An element belongs to the border of a subnet, if it is not connected directly to any element outside the subnet.

To illustrate we return to Fig. 1.9 in Chap. 1. The subset M, consisting of the elements $a?, h_a, a!, c_a$, for instance, is open. The *place* elements $a?, a!, c_a$ belong to the border, not however the transition element h_a, since it is connected only to elements *within* M. The subset $\{a?, h_a, c_a\}$, on the other hand, is not open, since here the *transition* element h_a belongs to the border.

The class of all open sets induces a so-called *topology* in the net, which then gives us a notion of continuity, generalized from the topic of real-valued functions.

The topology generated by place-bordered subnets features two particular properties defined by Petri. It is *primitive* and *elementary* (for an explanation of the terms see below). The mappings that are continuous with respect to this topology, he calls *net morphisms*. In this sense, the foldings above are in fact special morphisms. The notion of morphism does however also permit other aggregations, for instance the contraction of a place-bordered subnet to a single condition. However, the process foldings still remain the by far most important morphisms. The general theory of net topology was worked out in the 1970s by Petri's co-workers in the GMD; see for instance *Net Topology* [2] by César Fernández.

Hirzebruch and Pure Mathematics How he came to label his net topology as primitive and elementary, Petri later recalls as follows:

"When I came to the mathematics department in Bonn, Hirzebruch welcomed me most cordially. Knowing full well that pure mathematics professors looked down in disgust on computer professionals, I was moved by the warm reception, and soon I stole into the hall where Hirzebruch was lecturing on topology. I did not understand a single word; I had finished my studies of mathematics and had never heard of topology. I was frightened."

"Later, I saw in a glass case a wire model, standing for proof of a problem solution in eleven dimensions! A combinatorial insight! A wire toy for proof! I resolved immediately to study topology. Four years later, I wrote a paper of application for the Hausdorff award. (Hausdorff is the founder of modern topology.) I had no courage to go to Hirzebruch—he was world-famous even at that time. So I showed my paper to Hirzebruch's assistants. They turned me down completely, saying that my ideas were *primitive* and *elementary*. I have taken that message to heart. I did not rest until I could define 'primitive' and 'elementary' reasonably in formulas and could prove the theorem that net topologies are precisely the primitive and elementary topologies."

Petri calls a topology *primitive* if each singleton set is either open or closed, *elementary* if the class of closed sets also forms a topology. Here, as usual, a subset is closed if its complement is open.

Later Petri occupied himself with even more abstract structures, the so-called categories, in which the notion of morphism also plays a central role. In connection with a visit to the University of California Los Angeles, he is actually said to have read the abbreviation UCLA as the request: *U*se *C*ategory *LA*nguage.

Fig. 7.3 Petri with Möbius
strip (cf. also Fig. 2.1) and
self-intersecting planes

However, as became clear in subsequent years, it is indeed quite possible to
understand net theory without complex notions from topology and category theory.

One important role net topology *did* nonetheless play, namely in the mediation
between the unbounded "computing universe" developed by Konrad Zuse, and its
finite representation.

7.3   Petri and Zuse

Konrad Zuse (1910–1995) is one of the great pioneers of information processing
in Germany. He is the inventor and constructor of the first program-controlled
computer worldwide, the Z3 from the year 1941 (for which, as Petri once noted,
he was refused a patent, for "lack of inventive step"). To program the machine, Zuse
developed the first high-level programming language, called *Plankalkül*, between
1943 and 1945.

With his Plankalkül he later performed some work for the Gesellschaft für
Mathematik und Datenverarbeitung (GMD). In 1972 his report *Das Plankalkül* was
published in the GMD, where it later also appeared in an English translation titled
The Plankalkül. However, in the Anglo-Saxon world Zuse could not really catch on;
there others are regarded as inventors of computers and programming languages.

The Computing Universe In his later years Zuse turned his attention to the
theoretical foundations of computer technology. Like Petri he posed the question,
whether the natural laws of physics could not be directly incorporated into the design
of stable dependable computers. But in contrast to Petri, he arrived at the converse
idea, that it could be easier to express the nature of physics directly in the language
of computer engineers. If that would succeed, the world itself would appear as a
gigantic computer.

Fig. 7.4 Petri and Zuse 1975

As Petri recalls, Zuse in all earnest attempted this approach, much to the head-shaking disbelief of his contemporaries. He began with the design of "cellular automata". In this construction, space is conceived as a set of identical cells, which like small machines, exchange information with the neighboring cells, following a given behavior pattern. In 1969 he published his findings in the book *Rechnender Raum* (The Computing Universe).

Cooperation One day Zuse visited Petri in his office in the GMD and told him that he, Petri, had been recommended both as an ingenious theoretician and as a practically skilled manager of a large computer center. Zuse asked Petri to assist him in his investigation of the computing universe. This encounter developed into a fruitful cooperation.

Initially, Zuse advocated his approach vigorously, defending it against Petri's objections. As Petri later recalled, these were essentially the following:

First, the geometric arrangement of the cells in space induces a structure characterized by three distinguished directions. This, however, contradicts the fundamental cosmological assumption that space is *isotropic*. The *an*isotropy can then only be eliminated by the introduction of a "blurring factor", i.e., statistical noise, corresponding to random processes.

Secondly, Petri objected that in such a model the whole of physics would be "hidden" in each cell, so to speak as a non-analyzable DNA, without direct reference to physical entities.

Zuse and Nets As an alternative, Petri proposed a different modeling technique, namely the one of nets, which appealed so much to Zuse that he later wrote two books about it: *Petri-Netze aus der Sicht des Ingenieurs* (Petri Nets from the Viewpoint of an Engineer, 1980) and *Anwendungen von Petrinetzen* (Applications of Petri Nets, 1982). Petri succeeded in convincing him, that within the topological structure of nets, only finite nets are continuous, and thus able to provide a finitary image of a borderless universe.

They then agreed that the next practical step in the minimization of the switching elements would necessarily lead into the realm of quantum effects. They decided to leave it to the professional physicists to transcribe their insights into the extensive mathematical apparatus of quantum mechanics, a work for which the latter would probably need quite some time, "once they have started." Having reached this level, Petri and Zuse ended their joint activities.

Years later Petri recalls the cooperation with Zuse in a video clip (in German), available at the website of the Kurt-Pauli-Stiftung, a German foundation disseminating amongst others the ideas of Konrad Zuse: "The relationship was actually quite simple. He asked the questions, and I had to answer them. But there is an enormous difference, on what *intellectual level* the questions are asked, and Mister Zuse was able to ask questions on a level highest of all, and on the level of fundamental research."

A very personal memento from that time was a large abstract painting by Zuse that hung in Petri's office in the GMD until his retirement. According to Petri's son Tobias, the painting could be interpreted as a perennial with slender blossoms opening themselves to the light.

Non-Sequential Processes and Concurrency Theory

8

Contents

Processes run in given systems, and conversely, systems can be generated from processes by foldings. Information technological processes are by nature based on physical processes. The last remark provides ground for the question: Should it not be possible, beginning with the foundational properties of physical processes, to investigate how and why these can be interpreted as information-carrying processes; initially *without any reference at all to the concept of systems*, in which they may run? The renowned Dutch computer scientist Edsger Dijkstra once pointed out, that informatics is no more a science of computers than astronomy is a science of telescopes.

Under the heading *concurrency theory*, Petri dedicated a large part of his research activities precisely to that question, i.e., the bottom level in Fig. 7.1 in the last chapter. In order not to be influenced by preconceptions, he chooses a purely axiomatic approach. Some first indications of his ideas in this direction can be found already in the 1965 paper *Fundamentals of the Representation of Discrete Processes* [8]. To anticipate the central point in his findings: The axiomatic method again leads to a characterization, where the well known occurrence nets in level 1 of Fig. 7.1 turn out to be natural representations of processes. This of course once more confirms and reinforces the net-theoretical approach.

© Springer-Verlag Berlin Heidelberg 2015

E. Smith, *Carl Adam Petri*, DOI 10.1007/978-3-662-48093-9_8

8.1   Basic Concepts

In this chapter the presentation will occasionally have to be somewhat more formal. As a preparation, let us first recall, and slightly rephrase, the basic concepts of non-sequential processes as carriers of system runs.

A non-sequential process consists of an *acyclic* net of *occurrences of conditions C* and *events E*, called an *occurrence net*. The directed edges between conditions and events represent the *flow relation F*. The *directed paths* composed of edges, define a *partial order* "<" between the net elements. Hence also the notions and properties from the theory of partial orders carry over to occurrence nets.

Note however, that in concurrency theory some slightly different terms are commonly used. Two elements x, y, that are not ordered, are said to be *concurrent*. This is written as x co y. A *cut* is a maximal subset, in which any two elements are concurrent to each another. Dually, a *line* is a maximal totally ordered subset, i.e., in which $x < y$ or $y < x$ for any two distinct elements x, y, in other words, a *sequential subprocess*.

8.2   Signal Spaces

Petri begins, so to speak, completely from scratch, without any presuppositions. He bases his approach on the abstract concept of *signal*. By signal he understands everything that can be considered as physical carriers of information, as for example particles like electrons, frequency modulated AC voltage, hormones in biological systems, or even billiard balls. The essential point is, that signals *propagate* and *interact*. Informally, we call such a mesh of propagation and interaction of signals a *signal space*.

In the characterization of signal spaces, Petri resorts to studies by Hans Reichenbach, Kurt Lewin and Rudolf Carnap, who, in accordance with the results of special relativity theory, had formulated an axiomatic system of signal propagation in space and time in the 1920s. In fact, Petri considered Einstein as the actual inventor of the concurrency concept: "He called it causal independence, and defined it pragmatically like this: x is causally independent from y if and only if no signal departing from x can reach y and vice versa." The basic idea is illustrated in Fig. 8.1. The figure shows some signals and their interactions; p denotes an interaction

Fig. 8.1 Signal space

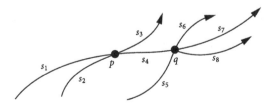

between the signals s_1 and s_2. The result propagates in form of the signals s_3 and s_4. At the point q, a further interaction between the effect of p (mediated via s_4) and another signal s_5 takes place.

From the finiteness of signal-velocity and -density, it follows that only a limited number of signals can take part in an interaction. Interactions between signals are *local* phenomena. Moreover, at the lowest level of abstraction considered here, we do not associate any form of "genidentity" (a term introduced by Kurt Lewin to describe an existential relationship underlying the genesis of an object from one moment to another) to signals, conserved across an interaction: In the example above, both s_3 and s_4 are considered to be different from s_1 and s_2. Similarly, the question, which one of the signals s_6, s_7 or s_8 is "new", created in the interaction q, has no meaning.

It is this general physical idea, that Petri takes up, for instance in his 1982 paper *State-Transition Structures in Physics and in Computation* [18], when he notes: "All physically possible computations can be described as embedded in physical processes."

Actually he can limit the class of possible processes further. As mentioned, local time reversibility is a central issue for him, such that he generally presupposes every interaction to have

(∗)   at least two causes and at least two effects.

He thus summarizes his approach as follows: "Two-particle interactions are sufficient to support well-organized information flow, including computation. We restrict our attention to the physical effect of a single and very simple structural type: that of the *interaction between two (or more) particles*."

Discrete Partial Orders What then should be required of a model of such signal spaces? It is obviously necessary that (1) the geometry of the individual signal chains, i.e., $s_1 p s_4 q s_6$, $s_5 q s_6$, etc., is adequately represented, but also that (2) the structure of the interactions is faithfully mirrored.

The resulting structure is a *discrete partial order*; a partial order, because in a relativistic sense one signal can only influence others in its postcone determined by the speed of light, and discrete, because the set of events is in fact discrete. How the signals behave *between* interactions, is not relevant to the interaction structure.

The fundamental relations thus turn out to be, on the one hand, "later" (in the sense of "causally influenced by"), when a signal or event lies in the postcone of the other, and, on the other hand, the complementary relation "coincident", when they can have no influence on one another. In a relativistic sense, coincidence replaces contemporaneity, however, in contrast to the latter, coincidence is *not transitive*.

For instance, in Fig. 8.1, s_5 is coincident to s_2, p, and s_3, whereas both p and s_3 are causally later than s_2.

8.3   Discrete Density

For a mathematical model of physical processes, the notion of *density* appears to be a natural requirement: For any two points in time, we assume that there is another one in between them.

Discrete orders, however, *cannot possibly* be dense in the classical sense. To this Petri replies that density as an abstract concept, in the sense of infinitary mathematics, is in fact not even necessary for the demands of information processing. For instance in *Non-Sequential Processes* [12] he notes: "Now let us recall the *purpose* of introducing the density postulates: to ensure that for every new, real observation a place can be found in an ordering scheme according to its relations to previously made observations. By requiring density we believe we can reserve a sufficiently number of such places."

According to Petri, the density requirement in a real-world application can be weakened to provide a number of values sufficient to represent the phenomena *actually occurring* in it.

As discrete counterparts to the classical density postulate, Petri suggests two requirements: N-density and K-density.

N-Density Effects between interactions are transmitted by signals. If two events are connected by a signal path, there must exist at least one signal-representation on that path. Considering the above mentioned requirement (∗), for the mathematical structure this can be translated into:

Between any two branching elements, there is always another third element.

He calls this postulate *N-density*. The term immediately becomes clear, when we look at it graphically. Figure 8.2 illustrates N-density by means of the signal space in Fig. 8.1: A representation containing p and q, must also contain an element in between, that stands for the mediating signal s_4.

N-density is a *local property* that "requires some intermediate element for well-defined purposes only" [18]. It is not a strong requirement; it is satisfied automatically in occurrence nets. Its purpose is mainly to show that, in the formal analysis of signal spaces, one is directly led to consider the model of occurrence nets.

K-Density In contrast to local N-density, the second requirement refers to the structure as a whole. As explained in *Non-Sequential Processes* [12], the idea is to generalize "the well known postulate from physics that every time sequence and

Fig. 8.2 N-density

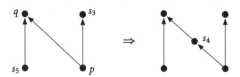

every space must have an element in common." Petri reformulates the postulate for processes as follows:

> We require that for every *case* and for every sequential process *p* it must be well defined how far this process has progressed. … This means however that every maximal totally ordered subset … must have an element in common with every cut.

He calls this property *K-density*. In *State-Transition Structures* [18] he summarizes: "The old physical postulate that every Cut represents a spatial distribution can now be written as 'every Cut cuts every Line'." The term "Cut" is again understood in the sense above, i.e., as a maximal subset, in which any two elements stand in the co-relation.

Unlike the letter N in "N-density", the K in "K-Density" does not refer to any graphical pattern. The K simply stands for the first letter in the Japanese word "*K*en", denoting certain *regional districts* in Japan, more precisely, 43 out of the 47 prefectures.

All occurrence nets are N-dense. All *finite* occurrence nets are also K-dense. As a requirement, K-density is therefore only relevant for infinite occurrence nets. In *Non-Sequential Processes*, Petri notes that also in infinite nets the "requirement can only be violated if an infinite number of sequential processes have to wait for each other, i.e., a situation of no practical interest."

Petri concludes that, adhering to the relevant fundamental physical laws in a strict sense, it is precisely the K-dense occurrence nets that present themselves as natural carriers of information processing.

The Continuum Net   Classically in modeling, events in physical processes are associated with time points *t* from the real-valued continuum $\mathbb{R}$. This approach however deals *only with events*, and not with the *states between* the events.

Fig. 8.3   Petri at the University of Kyoto. 43 out of the 47 prefectures of Japan are called *Ken*. Kyoto is one of the four exceptions

It is however straightforward to augment a representation containing the events t_1 and t_2, $t_1 < t_2$, by a state s that begins with the occurrence t_1 and ends with t_2; it suffices to regard s as the set of time points between the event occurrences. Mathematically, then s becomes the *open interval* $\{t \in \mathbb{R} \mid t_1 < t < t_2\}$, often also written in the form $s = (t_1, t_2)$. From such events t and states s it is then again easy to construct occurrence nets by building the flow relation F from components of the form $t_1 \, F \, s$ or $s \, F \, t_2$ for $s = (t_1, t_2)$.

Example

Figure 8.4 shows a segment of the course of a year, say the year 2015.

The events $t_1, t_2, t_3 \in \mathbb{R}$ denote respectively the beginning of spring, summer and autumn. The state a denotes spring. It can be conceived as the set of time points between the beginning and end of spring, i.e., the open interval $a = (t_1, t_2)$. By the same token, the state b (summer) corresponds the interval $b = (t_2, t_3)$. The state $c = (t_1, t_3)$, finally, holds during the two "warm" seasons spring *and* summer. In net terms, the flow relation $t_1 \, F \, a$ holds between the event "beginning of spring" t_1 and the spring a. Similarly, the flow relation F also holds for the remaining net elements connected by arrows.

But also the converse is true: For *every* occurrence net of practical interest (or equivalently: signal space) we can derive a representation with events from $\mathbb{R}$ and states consisting of suitably chosen open intervals.

Example

For the signal space in Fig. 8.1 we can for example define such a net as follows: It consists of the events $p = 1, q = 2$ and the states $s_1 = (0, 1), s_2 = (1/2, 1), s_3 = (1, 3), s_4 = (1, 4), s_5 = (1/2, 2)$ and $s_i = (2, i)$ for $i = 6, 7, 8$. The flow relation between the net elements is then induced automatically.

Definition The most comprehensive net definable in this way, is the *continuum net*. It consists of the set of events $E = \mathbb{R}$, and the set of states B of all *bounded open intervals* b in the real numbers $\mathbb{R}$, i.e., the set of all b of the form $b = (t_1, t_2)$ for

Fig. 8.4 A segment of the Four Seasons net

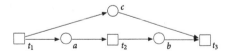

$t_1 < t_2 \in \mathbb{R}$. The flow relation F between events t and states b is then given by

$$t \, F \, b \Leftrightarrow t = t_1 \text{ and } b \, F \, t \Leftrightarrow t = t_2 \text{ for } b = (t_1, t_2).$$

In *Non-Sequential Processes* [11], Petri emphasizes the relevance of the continuum net in form of a thesis as follows:

1. Every net representation of a real process is isomorphic to a K-dense subnet of the net associated with the continuum.
2. In general, real processes are non-sequential. Their conditions can be chosen as causal connections of events.
3. Not all time intervals can correspond to causal connections.

Item 1 is the actual "representation theorem": All occurrence nets of practical interest, i.e., K-dense ones, can be represented by events from $\mathbb{R}$, and states defined by open intervals in $\mathbb{R}$.

Item 2 states that a causal connection between events t_1, t_2 can for instance be mediated by the condition given by the open interval (t_1, t_2).

Item 3, finally, states that the interval (t_1, t_2) may not be included, when there is *no causal connection* between the events t_1 and t_2.

Example

It may be useful to clarify items 2 and 3 a little bit more with another example. Figure 8.5 shows an occurrence net that is to be interpreted as a subnet of the continuum net.

This can be achieved for instance by choosing $t_i = i$ for $i = 1, 2, 3, 4$, which then also determines $b = (1, 2)$ and $c = (2, 3)$.

For a we choose any interval of the form $a = (t, 1)$ with $t < 1$, for d any one of the form $d = (3, t)$ with $3 < t$, and likewise for e and f.

Between t_1 and t_3 there is a causal connection, which however is not established by a directly mediating condition (which would have to be the interval $(1, 3)$). The event t_4 is causally independent from the other t_i, because none of the possible "causality-mediating" intervals $(i, 4)$ for $i = 1, 2, 3$, occur in the net.

The Continuum Net is not K-Dense It is interesting to note that it is actually the classical density of the real numbers, that guarantees that all processes of practical interest, represented by K-dense occurrence nets, can be embedded into the continuum net. However, on the other hand, this same density property has

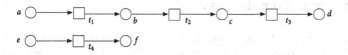

Fig. 8.5 Subnet of the continuum net

a remarkable—almost paradoxical—consequence, namely that the continuum net itself is *not* K-dense. Since a detailed proof of this fact does not seem to be available in the literature, we take the opportunity to give one here.

Theorem (Petri [11]). *The continuum net is not K-dense.*

Proof It is clear that the set $X := E \cup B$ is not totally ordered, because, for $b, c \in B, t \in E$ we have:

(1)   (*a*) b co $c \Leftrightarrow b \cap c \neq \emptyset$ and   (*b*) t co $b \Leftrightarrow t \in b$.

On the other hand, the set $E = \mathbb{R}$ is obviously a totally ordered subset of X. Moreover, it is also maximal, hence a *line*, since for every $b \in B$ we have b co t for $t \in b$.

To prove the theorem it therefore suffices to find a cut in X, that has no element in common with the line E. To this end we pick an arbitrary t in E, and define

(2) $C_t := C_{t,1} \cup C_{t,2}$ with $C_{t,1} := \{b \in B \mid b \, F \, t\}, \quad C_{t,2} := \{b \in B \mid t \in b\}.$

We show that this set C_t is as required.

First we observe that $C_t \cap E = \emptyset$ is obvious by definition of C_t.

Next we show that any two elements in C_t are unordered. For two $b, c \in C_{t,1}$, the element t is the common upper interval limit. Hence we must have $b \subseteq c$ or $c \subseteq b$, in any case $b \cap c \neq \emptyset$, and thus b co c by (1*a*). For two $b, c \in C_{t,2}$ we have $t \in b$ and $t \in c$, thus also in this case b co c.

We come to the mixed case $b \in C_{t,1}, c \in C_{t,2}$. Note that now the upper limit of the interval b is contained in the interval c. Since b and c are open intervals, they must therefore have a non-empty intersection, hence also in this case we have b co c.

It remains to show that C_t is maximal, i.e., that every element $x \in X$ that is co to every element in C_t, must itself belong to C_t. So, let x be such an element. Note that $x \notin E$: t itself is by definition greater than any $b \in C_{t,1}$; and a $t' \neq t$ cannot be contained in every $b \in C_{t,2}$.

It follows that $x = c$ for some $c \in B$. If $t \in c$, there is nothing to show, since then $c \in C_{t,2} \subseteq C_t$ by definition. We still have to consider the final case that $t \notin c$. But then, since c intersects every $b \in C_{t,1}$, t must actually be the upper limit also of c, hence $c \in C_{t,1} \subseteq C_t$, as desired.

We have thus shown that C_t is indeed maximal, which now concludes the proof.□

8.4   Discrete Completeness

In the preceding sections we followed Petri's argument that it is neither necessary nor useful to base the description of information processing procedures on classical density requirements: It is sufficient to consider K-dense occurrence nets.

But then another question arises, namely how to model the *perception of continuous processes*, the *sensation of movement*, within such a discrete approach? We *perceive* movement as an immediate basic phenomenon; the resolution into a set of single points is a subsequent act of the mind.

The ancient Greek philosopher Zeno of Elea, contemplating the movement of a flying arrow, asked himself how he could perceive the arrow's motion, even though in any moment of its flight, it occupies an exactly defined position, in which it is at rest, because it is not possible to move within a given position. Thus, since the arrow is at rest in *every instant*, and time is composed of instants, it must remain motionless *throughout*.

In classical physics, the possibility of motion is derived from the concept of *infinitely short* duration of stay in each point, and formulated mathematically with the help of limits of converging sequences. In the corresponding mathematical model, it is necessary to require so-called *completeness* in addition to density.

As in density, Petri now reformulates the concept of completeness, in order to reduce it to its actual purpose and content, namely the requirements of real discrete information collection and processing.

In *Concurrency and Continuity* [19] he notes—in the spirit of Zeno: "The movement of a falling apple is called continuous. We are quite certain that we can catch the falling apple in our hands at different heights above the ground, and indeed at any height we choose between the different heights we might have attained. Also, by taking flashlight snapshots we can show the falling apple in different positions by freely choosing some of the various possible times for taking the snapshots. We are not able to count the number of the possibilities which are open to our choice, and thus we are led to think that between two chosen different possibilities there is always another one."

In addition to the notion of density discussed above, here the notion of a continuous motion within a process comes into focus. What Petri wants, is to consolidate the intuitive sensation of continuity on the one hand, and the discrete articulation of actual observations on the other.

Dedekind Cuts To this end, in his 1978 paper *Concurrency as a Basis of Systems Thinking* [16], Petri had already suggested a generalized definition of completeness for partial orders, based on the work of Richard Dedekind. This first approach turned out to be somewhat unintuitive and not immediately accessible. It is discussed extensively in the textbook *Nonsequential Processes: A Petri Net View* [1] by Eike Best and César Fernández, two of Petri's co-workers for many years in the GMD.

In a *totally ordered* set X, a *Dedekind cut* consists of a pair of disjoint sets A and B, such that $X = A \cup B$, and for $a \in A$, $b \in B$ always $a < b$. The order is then called *(Dedekind) complete*, if for every Dedekind cut (A, B), either A has a greatest or B a least element. As known, the set of real numbers $\mathbb{R}$ is complete in this sense, the rational numbers $\mathbb{Q}$ however not.

In his earlier proposal, Petri takes up the notion of the *bipartition* of the set X, and attempts to generalize it to partial orders. Later it turned out that a more suitable approach is to base the generalization to partial orders on a somewhat different formulation, which in *total* orders is *equivalent* to Dedekind's original one.

It states: A Dedekind cut is a pair (A, B) of subsets, such that A consists precisely of all elements that are smaller than every element in B, and conversely B of all elements that are larger than every element in A.

In this definition there is no mention of a disjoint partition of the base set. In total orders, bipartition follows as a consequence, in partial orders however not necessarily.

Cone Pairs In 1987 Petri takes up this second variant in *Concurrency and Continuity* [19]. A Dedekind cut in a partial order does then no longer induce a partition, but rather a *cone pair* in the sense that the precone A together with an element a also contains all predecessors $a' < a$, and conversely the postcone B together with a b contains the set of all successors $b' > b$. The definition of Dedekind completeness itself is carried over literally.

Example

We illustrate the idea by means of the partial order in Fig. 8.6. (The reason for the representation of the elements as circles and squares will become clear below.)

A Dedekind cut in this order is for instance given by the pair (A, B) with $A = \{s_1, s_2\}$, and B consisting of the all remaining elements *except* s_5. In particular we note that it does not form a partition of the base set. The cones meet with a gap: they are both truncated in the sense that neither A has a unique greatest element, nor B a least element. The order is thus not Dedekind complete. If we now fill the gap with an additional element p, such that $s_1, s_2 < p < s_3, s_4$, then (A, B) is no longer a Dedekind cut, but for instance $(A \cup \{p\}, B)$ *is*. As a result of this gap-filling, the extended structure becomes Dedekind complete. Moreover, note that it now actually corresponds to the occurrence net of the interaction structure in the signal space in Fig. 8.1.

Complete and Dense Occurrence Nets As in the example, it is also true in general, that *every occurrence net*, in which each event has a least two preconditions and two postconditions, is *complete* in this sense.

The above correspondence between single point gap filling, on the one hand, and elements of occurrence nets, on the other, can be carried even further. A deletion of an *event* from an occurrence net results in a structure that is still N-dense, but leaves a (Dedekind) gap, which in turn indicates where to insert a missing T-element.

Fig. 8.6 Partially ordered set. Not Dedekind complete

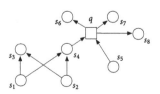

Conversely, a deletion of a *condition* results in a structure that is no longer N-dense, and the gap indicates the missing *S*-element. In both cases, the position of the required completion is uniquely determined.

In *Concurrency and Continuity*, Petri summarizes the requirements of density and completeness above as follows: "As a consequence, physical signaling structures can be modeled as a combinatorial (in this sense 'discrete') continuum."

Petri was well aware that the notion of 'discrete continuum' might raise an eyebrow. Perhaps this is why he preferred the term 'combinatorial'. However what he meant was, nonetheless, that the actual perception and experience of continuous processes can be described in terms of discrete structures.

8.5   Combinatorial Mathematics

Another reason why Petri occupied himself so thoroughly with the concepts behind density and completeness was because he had a vision of a combinatorial mathematics, based directly on the requirements of practical observations and activities. In this way he intended to avoid a detour into inappropriate classical infinitary mathematics.

In fact, every collection of actual observations—such as the gathering of measurement values—is finite, i.e., exists as a discrete set (more on this in Chap. 10). Classically, this set is then embedded into the real numbers, where it can be processed using e.g. differential equations. For a numerical computation of these equations, in turn, finite approximations are applied, which however in general have no relationship whatsoever with the finiteness of the original data. For this reason, Petri saw a long-term objective in the development of an appropriate combinatorial mathematics, that works directly with the observations, without the need for any twofold "digital-analog, analog-digital" conversion.

There is clearly still a long way from the discussions above to a comprehensive combinatorial continuity-mathematics. However it is undoubtedly to Petri's merit to have reduced the essential modeling questions to the discrete nature of their objects, without any presupposed idealistic hypotheses—in accordance with Newton's "non fingo", that Petri had already followed in his asynchronous approach to automata theory.

Communication Disciplines

9

Contents

As mentioned before, in the 1980s Petri took part in the early development of the ARPA-net, the predecessor of the Internet. In view of the novelty of such a network, by which computers were to be connected on a large scale for the first time, new questions and problems naturally arose, from purely technical ones like physical connection issues, to questions concerning the handling of the net by the individuals involved. Through networking, computers now indeed turned into *media of communication*, so that the notion *communication with automata* in Petri's thesis finally *could* be understood in the sense of communication *by means of* automata, regardless of whether this was intended from the beginning or not.

Petri was able to contribute to the solution of various practical problems in the ARPA-net. His friend and colleague Lu Ruqian reported—with a Chinese love of numerical detail—that Petri found about 1300 design- and programming-errors from 450 problem classes through analysis of the documentation of the test runs, which he then divided into twelve large groups.

9.1   The Map

Even though it is difficult to confirm the numerical particulars years afterwards, here clearly lay one of the origins for the identification of twelve subareas of communication, which Petri documented in the 1979 publication *Kommunikationsdisziplinen* (Communication Disciplines) [17], a preliminary version in English can be found in [14]. His division into problem groups is shown in Fig. 9.2. Petri stresses the point however, that this division should be treated with every caution, "whether the

© Springer-Verlag Berlin Heidelberg 2015
E. Smith, *Carl Adam Petri*, DOI 10.1007/978-3-662-48093-9_9

Fig. 9.1 Petri in "largely
unknown territory" [17]

partition is complete and whether it is appropriate," since—as he adds: "We enter
largely unknown territory and cannot be surprised if our preliminary map requires
correction."

The twelve subareas, which he understands as different communication "disciplines", are listed in the order of presumed difficulty for a complete mathematical
understanding, as it appeared to the parties involved in the ARPA-project at that
time.

Petri remarks on the classification as follows: "We could be tempted to speak of
the 'low-level' disciplines 1–7, which are apparently of purely technical nature, and
the 'high-level' disciplines 8–12, which appear only to refer to mental processes,
legal questions and organization." He stresses however that such a strict delineation
of concepts is hardly definitive.

Fig. 9.2 Communication
disciplines

1 Synchronization	2 Identification
3 Addressing	4 Naming
5 Copying	6 Cancellation
7 Formatting	8 Modeling
9 Authorization	10 Valuation
11 Delegation	12 Reorganization

9.2   The Disciplines

1. By *synchronization* Petri understands "in general all coordination of activities in temporal respect, whether they are of technical or organizational nature." As an example of how a mathematical theory for the treatment of the individual disciplines could be imagined, he discusses synchronization by means of the previously mentioned synchronic distance in Petri nets.
2. As examples in the discipline of *identification* he considers the question, "when an electric potential denotes a determined logical value," but also the "proof of administrative responsibility of instances for certain actions."
3. By *addressing* he understands the "description of paths and path-systems by message components."
4. A typical question belonging to the discipline of *naming*, is for example: "How can the incompatible naming conventions for files in COBOL and FORTRAN be understood as special forms of a common single naming discipline?"
5. A central problem in *copying* is the change of a message's pragmatic status when it is copied. By copying, the message for example may lose its status as a document.
6. A typical question in the discipline of *cancellation* could be: "Is there a difference between how error correction should be dealt with, versus the case that an instance wishes to withdraw a decision because of 'change of mind'?"
7. *Formatting* deals with rules for the structure of messages. Moreover, Petri sees the formatting discipline as highly relevant in the "context of development of abstraction models in the discipline of modeling."
8. With respect to *modeling* he points out: "The origin of model-schemes is usually a historic process, evolving without conscious methodological techniques." As an example he mentions that models that have proved to be useful in one field, are carried over to other fields without necessary care, as for instance "the use of real numbers to denote real measurement values, even though the validity of the axioms for real numbers cannot be established for measurement values."

    In the next chapter we shall discuss the problems of measurement values in detail.
9. *Authorization* is concerned with technical access rights, but in the context of personal instances also with authority to issue directives, and their control. This discipline should include a way of thinking that "derives ethical restrictions and technical control from common principles." Petri however points out that this appears to be an issue still a long way away.
10. *Valuation*: Information is to be considered as a resource, "as something that is distributed unevenly, and therefore scarce in some places. The subject of a theory would thus in the first place have to be the exchangeability of resources that are different *by nature* (time, memory, energy, tools, programs, information, etc.)."
11. *Delegation* is the transfer of tasks from one instance to another. An important question is where the responsibility associated with the task is to be located. In

delegation to a non-personal instance, for example, the responsibility cannot be delegated along with the task.

12. By *reorganization*, finally, Petri understands "Action schemes for the organizational alteration of a system."

Here another source for the communication disciplines becomes clear, namely the meta-communication in the extension of automata, which Petri discussed in his thesis.

An essential requirement in the discipline of reorganization, in which traces of the thesis are particularly easy to recognize, is: "Such schemes are expediently designed together with the system to be developed, and should be considered as part of the system proper. Such action schemes especially show why sequential schemes often do not suffice; beyond the dimension of sequence, they in general require the dimensions of alternative and concurrency."

9.3   Channel-Instance Nets

For the formal treatment of the "higher" disciplines, Petri proposes the use of correspondingly more abstract Petri nets. These also consist of two sorts of elements (or *functional units*), namely *channels* to represent messages, and *instances* for their processing.

Graphically, the functional elements are again denoted by *S*-and *T*-elements. The edges in such a channel-instance net is not necessarily directed, however. If for example we only wish to show that instances take part in activities, it is not always necessary to specify the direction of information flow, such as suggested by the token game in condition-event systems. As an example of undirected nets, Petri mentions the so-called *role-activity nets* introduced by Anatol Holt.

Channel-instance nets are also useful for specification purposes in the early phase of modeling, which can then be refined to a representation based on condition-event systems. For instance, channel-instance nets can be used to formulate the meta-communication in the extension of automata, that Petri discussed in his PhD thesis (e.g., how to organize additional tape supply), which thus may at a later stage be translated into the unified language of basic nets.

Theory of Measurement

10

Contents

According to Petri, computer science is not limited to information *processing*, but should also include questions of its *acquisition* and *usage*. His view on the overall context is explained in detail for instance in *Kommunikationsdisziplinen* [17], discussed in the last chapter. In particular, the handling of *continuous quantities* and their *discrete representation* plays a prominent role. Beginning with his experiences as artillery assistant in his youth, he became especially interested in the fundamental concepts behind the gathering of information in the context of measurement.

Later he developed this into an empirically based theory of measurement. In 1977 he presented an outline in a lecture *Modeling as a Communication Discipline* [15]. Although the material does not appear to exist in a published form (the reference only contains a short abstract), the ideas are of crucial importance for the understanding of Petri's work and thoughts. We shall therefore discuss them in some detail. The presentation is based on elaborations by this author, that he made in his PhD thesis under Petri's supervision.

© Springer-Verlag Berlin Heidelberg 2015
E. Smith, *Carl Adam Petri*, DOI 10.1007/978-3-662-48093-9_10

10.1   Classical View

Classically, measurement is conceived as the assignment of appropriate real numbers. According to a German DIN standard *Grundbegriffe der Messtechnik* (Basic Concepts of Measuring Technique, DIN is an abbreviation for Deutsche Industrie Norm, comparable to the ASA standard in the United States), measurement consists of "establishing the special value of a physical quantity as a multiple of a unity or reference value. The objective of a measuring is to establish the true value of a quantity."

In this view, numerical values are inherent properties of objects. However, as the DIN standard also points out: "Due to practical limitations in the measuring process, deviations will occur. They are the reason, why it is not possible to establish the true value x_W."

In mathematical terms, the objective of measurement is explained for instance in the text book *Axiomatische Grundlagen einer allgemeinen Theorie des Messens* (Axiomatic Foundations of a General Theory of Measurement) by the Austrian mathematician and statistician Johann Pfanzagl: "The general aim of a measurement is to associate a real number to every element of a given set M, such that relations between the value numbers enable conclusions to be drawn about the corresponding relations between the elements of the set M."

To summarize: In these approaches, observation and measurement consist in establishing and reporting the true value more or less accurately. Comparison between different objects can then be drawn by considering the associated numbers. Measurement—i.e., the assignment of such numbers—seems to precede comparison conceptually as well as practically.

10.2   Petri's Approach

But then the question immediately arises, if the assumption of inherent real-valued magnitudes is actually necessary or appropriate, when it inevitably implies that every assessment of measurement values is faulty, and as a consequence every discrete representation of continuous magnitudes must be considered merely as an approximation.

We arrive at a more comprehensive understanding of measuring, when the technical procedure is not considered in isolation, but rather within the *pragmatic whole* of the application context. In this view, the real objective of measurement is to provide grounds for *decisions between various alternative courses of action* in real-world situations. Thus the pragmatically relevant *information content* of a measuring will depend on the number of possible alternatives. This number, however, is clearly always bounded. In this sense, every measurement generates a *finite* amount of information.

In contrast to the classical approaches discussed above, Petri advocates the view, that in dealing with actual continuous magnitudes it is more appropriate to avoid the idealistic detour via real numbers. Rather, one should acknowledge the finiteness of the information content in measurements already in the construction of the theory.

Also in this context Petri suggests using "elements of a suitable chosen structure" instead of real numbers. Moreover he sees the need to develop the foundations of measuring, based on an understanding of the activity of direct *actual comparison*. He proposes to start from the empirical observations themselves, i.e. to use the relation *observably larger* and its complement *empirical indifference*, also called *near-equality* or *indiscernibility*.

10.3 Empirical Indifference

The idea behind the notion of empirical indifference can be illustrated in a weight comparison of objects x, y with a beam balance. If the pan containing y sinks after x and y have been put in the pans, it is obvious that y is heavier than x. In this way the balance induces an order "y is observationally heavier than x". We write this as $x < y$. It is however also possible that x and y balance each other out. In this case we write $x \approx y$. The relation $\approx$ denotes the empirical indifference "equilibrium determined by means of the balance".

The crucial question now becomes: If x balances out y, and y balances out z, will then also x balance out z? In formulas: If $x \approx y$ and $y \approx z$, will then $x \approx z$ also hold? Mathematically: Is the relation $\approx$ transitive?

In fact, this question has been answered to the affirmative by many authors, for instance by the philosopher Rudolf Carnap, already mentioned in the context of signal structures, in his book *Einführung in die Philosophie der Naturwissenschaften* (Introduction to the Philosophy of Natural Sciences). Petri rejects this assumption vehemently. He argues that transitivity of $\approx$ cannot be justified *empirically*; on the contrary, it can be refuted by counter examples.

Example

The following example illustrates the idea. From a long wooden stick, copies of a given stick x_0 are to be cut off. To this end, the original x_0 is placed along the long stick and its length marked, say using a pencil, and then the copy x_1 is cut off at the mark, *and so forth*. Now the question is, what precisely do we mean by "and so forth"? Do we take the copy to determine the next copy, and then the copy of the copy? In this case we get

(1) $$x_0 \approx x_1, x_1 \approx x_2, x_2 \approx x_3, \ldots$$

Or do we rather always use the same original x_0? *Then* we get

(2) $$x_0 \approx x_1, x_0 \approx x_2, x_0 \approx x_3, \ldots$$

Everybody who has tried method (1), knows that it results in a significant difference to the original, after only a few steps. In method (2), on the other hand, the desired "near-equality" can be retained. Petri concludes that the empirical indifference is in fact *not* transitive.

The Bounds to Non-Transitivity   But now, if empirical indifference is not transitive, are there no reliable assertions that can be made? Fortunately this is not so, in fact very tight bounds for the deviation from transitivity can be given. Let us consider the task of sorting a set of wooden sticks according to length, by direct comparison. Assume a person P has not observed any difference between the sticks x and y, but noted that v is clearly shorter than x, and that y is shorter than z. Then it is plausible to assume that, in direct comparison, P will note that v is clearly shorter than z. Formulated as a postulate:

(P1)   If $v < x, x \approx y$ and $y < z$, then $v < z$.

The postulate is illustrated graphically in Fig. 10.1.

Moreover, the following also seems plausible: If P has observed that x is shorter than y, and y shorter than z, then it is not to be expected that there is a stick that is indiscernible from both x and z. We can therefore assume that the following postulate holds:

(P2)   If $x < y < z$, then there is no v, such that $x \approx v \approx z$.

(P2) is illustrated in Fig. 10.2.

The postulates (P1) and (P2), on one hand, reflect the fact that empirical indifference is in general *not* transitive, but on the other hand, determine very *tight bounds* for the *deviation* from transitivity. These bounds are of seminal significance for Petri's approach. It may therefore be useful to give yet another illustration.

Example

Assume we have a beam balance, with which we can clearly distinguish sets of rice grains, when they differ by ≥ 9 grains. If however the difference is ≤ 6 grains, we can assume that the two pans balance each other out.

Fig. 10.1 (P1). Dotted lines denote $\approx$

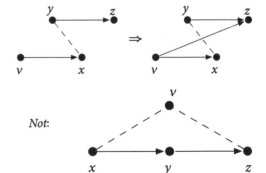

Fig. 10.2 (P2). Dotted lines denote $\approx$

Consider the situation in (P1). The variables now directly denote the number of grains. The arrow, say, from v to x then means that $v + 9 \leq x$, and similarly for the other arrows. The dotted line between x and y signifies $|x - y| \leq 6$. It is trivial to check that in this case $v + 12 \leq z$, such that the postulate (P1) holds for the measuring arrangement.

Moreover, (P2) also holds, because obviously $x + 18 \leq z$, hence there can be no v, such that $|v - x| \leq 6$ and $|z - v| \leq 6$.

Remark For the more mathematically inclined reader, we remark that the empirical relations above are not of the 'if and only if'-type. If the difference is, say, 7 grains, sometimes a user may judge there to be a balance, and other times to be an imbalance. Indeed, there is also in general no way to determine the exact borderline between certainty and uncertainty.

In contrast to Carnap, other theorists do recognize that the relation of directly observable empirical indifference is not transitive, but claim that there is a genuine transitive equivalence *underlying* it, that can be regained from the imprecise measurements by a posteriori deductions.

A prominent representative of this view is the US American mathematician and founder of cybernetics, Norbert Wiener. In an article *A new Theory of Measurement. A Study in the Logic of Mathematics* he discusses the issue in general terms, and arrives at—translated into the example of the beam balance—the following conclusion: If we observe that $x \approx y < z$ or $x < y \approx z$, then in both cases, z is in reality heavier than x. The complement relation to this "in reality heavier than"-order then, in fact, turns out to be transitive, i.e., an equivalence relation, like true equality.

However, Petri also vehemently rejects this approach, pointing out that it is only a purely *formal* construction, from which no increased accuracy beyond mere observation can be deduced.

10.4 Reference Scales

In the context of empirical comparisons, the construction of *measurement scales* consists of providing a set of distinguished comparison *references*. A balance on a vegetable market, for instance, is commonly equipped with a matching set of weights. Two characteristic parameters for such reference sets are *resolution* and *size*. The set should consist of a number of elements that permit measurement as "precise" as possible, i.e., should provide an acceptable resolution power. On the other hand, to be manageable, the number of elements should be as few as possible.

Resolution We begin with resolution: A set of references is clearly of sufficiently fine granularity, when any object to be compared (within a given range) will be empirically indiscernible to one of the elements of the set.

It is not immediately clear how such a resolution can be warranted by "endogenous" properties of a reference set, since it refers to *every conceivable* comparison operation.

Under the assumption of postulate (P2), this is, however, very easy. Then it is only necessary that the following postulate (P3) also holds:

(P3)   For any two direct successor objects x, y *within* the reference set, there is a third one z within the set, such that $x \approx z \approx y$.

Here y is a *direct successor* of x, if $x < y$ and there is no element v in the set, such that $x < v < y$. Formally this is often also written as $x \lessdot y$.

Example

To illustrate (P3) we return once more to the example with the beam balance in the last section above.

(1) We first consider a set Z of reference weights z_i, each consisting of $20i$ rice grains. This set is too coarse. For $v = 30$ grains we get $z_1 < v < z_2$ (where again $<$ stands for the relation "observably heavier"). In this arrangement, there is no reference weight that balances out v.

(2) To contrast, let Y be the set of all weights y_i consisting of $10i$ grains. Now it *is* true that *every* set v of grains to be compared, will balance out at least one of the references y_i. For $v = 17$ grains, say, we get $v \approx y_2$, for $v = 16$ even $y_1 \approx v \approx y_2$.

However, note that is not possible to differentiate between the situations (1) and (2) by inspecting the reference sets themselves; in both cases we always have $z_i < z_{i+1}$, respectively $y_i < y_{i+1}$.

(3) The situation changes however, when we consider the reference set X of weights x_i with $5i$ grains each. Now we get

$$x_i \lessdot x_{i+2} \quad \text{and} \quad x_i \approx x_{i+1} \approx x_{i+2}$$

within the set X. And that is exactly what is required by postulate (P3). By inspection of the reference set itself, we can thus guarantee that any weight to be compared, will always be balanced out by at least one of the references.

Coherence A consequence of (P3) is, that the *whole* of the reference set can be traversed in a sequence

$$x_1 \approx x_2 \approx \cdots \approx x_n$$

of indifference steps. Petri calls this property *coherence*. He sees it as an important general principle. Coherence is among others a necessary precondition for the sensation of continuity. A motion picture only appears continuous, because there is no noticeable gap between the single images, which means that the set of images is coherent in the sense above.

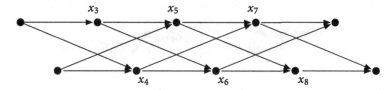

Fig. 10.3 Minimal reference scale. Arrows denote $<$, no connection: $\approx$

In the same context, also the question of Zeno of Elea could be discussed, why a single falling grain of millet makes no sound, a sack of millet however does.

Size A last question remains: What is the *minimal size* of a reference set that satisfies (P3). This, however, is also straightforward:

A reference set, for which (P3) holds, is minimal if every subset consisting of mutually indiscernible references has exactly *two* elements. Translated into the language of partial orders, the requirement becomes: Every *cut* consists of exactly two elements. Figure 10.3 shows a segment of such a "thin" reference scale. Examples of cuts are $\{x_3, x_4\}$, $\{x_4, x_5\}$, and so forth.

Example

We return to the example with the rice grains above. The set X in (3) has exactly the desired form. The relationships in Fig. 10.3 are immediately verified for the $x_i \in X$.

10.5 Petri's Double Scale

We can illustrate the empirical indifference and corresponding measurement scales by considering the somewhat less abstract example of a real analog-to-digital converter. Such a device converts a continuous physical quantity to a digital number, based on the reading of some kind of sensor. An often used type is an encoder, where a set of brushes passes over alternate segments of conducting or insulating material.

Since the physical brushes (or sensors, in general) are of finite width, the device will unavoidably run into situations where a value cannot be determined unambiguously. Technically, this can be accommodated for by sampling the analog data with *two displaced* sensors, such that in any circumstance at least *one* of the value readings is unambiguous.

Equivalently one can of course consider *displaced scales* instead of displaced sensors. Petri usually illustrated the idea behind such a construction with the *double scale* in Fig. 10.4.

The pointer in position A does not permit one to read off an unambiguous value on the upper scale. In such cases it is however *by construction* always possible to read off a *definite* value on the lower scale (here 4). If we therefore restrict ourselves to unambiguous values i, we obtain unequivocal order statements of the form $i < i + 2$, but we can in general not conclude $i < i + 1$ or $i + 1 < i + 2$.

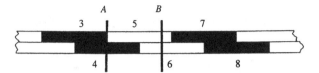

Fig. 10.4 Petri's double scale

This does not contradict the fact that in *many* cases (but not in all) it nonetheless *is* possible to read off an unambiguous value on both scales, here for instance the values 5 and 6 in position *B*. We denote such a value reading as 5&6.

Anyone who still has a strip of negatives of a 35mm film can, by the way, ascertain that these strips also use a similar double scale to uniquely label the photos.

Thus, if reading errors of size ±1 are unavoidable in principle, the requirement for a measurement scale must be that such an error is *tolerable in a given context*. For example, an optician needs a finer resolution than a carpenter.

A scale can be considered sufficiently precise, if the empirical order relation < can be overlapped by a sequence of near-equalities. In the example above, we have for instance 3 < 5 and 3 ≈ 4 ≈ 5. It is not difficult to recognize that the order structure in this case corresponds exactly to the requirements derived theoretically in the preceding sections. The postulates (P1) and (P2) are immediate; and also the size and resolution of the scale correspond to what was required there.

10.6   Conflict and Confusion

Interestingly, in the description of measurement processes, we encounter the same phenomena *conflict* and *confusion* as already observed in elementary nets. In the double scale in Fig. 10.4, it can be argued that pointer position *A* indicates a *reading conflict* between the values 3 and 5. When, conversely, a value 4 is reported, it is a posteriori not possible to determine whether the report of the value 4 was due to a deliberate change of scale, because of an ambiguous value on the upper scale, or not—i.e., a typical situation of confusion.

In philosophy, this phenomenon is known as the problem of *Buridan's ass*, which stands in the middle between two equally large hay piles, and then starves because it cannot decide which one to eat first. If the ass however *does* turn to one of them, it will a posteriori not be possible to determine whether that pile *was* closer, or if the ass had *resolved a conflict* through a decision.

10.7   The Two Roots of Concurrency Theory

If we compare Petri's development of non-sequential processes with his approach to measurement theory, it is striking that both are founded on discrete order structures, where the complementary relation *non-order* is of crucial significance. Both the concurrency relation co, and the empirical indifference $\approx$ are non-transitive. Indeed, Petri used to point out that both approaches are different manifestations of the same theory—i.e., concurrency theory. (By the way: we can now complete our explanation of the relations in Fig. 1.6 in Chap. 1: The fourth relation *sm* on the bottom right stands for *similarity* or *likeness*.)

In non-sequential processes, the notions of K-density and Dedekind completeness stand central. In measurement, the focus is on controlled deviation from transitivity in empirical indifference, and continuity in the sense of reachability in indifference steps.

Apparently this points more to differences than to commonalities, but in fact the interrelationship between the two questions is quite close.

10.8   Complete K-Dense Measurement Scales

Let us take a closer look at the scale in Fig. 10.3. It is clearly not K-dense, since e.g. a line that contains x_4 and x_7 has no element in common with the cut $\{x_5, x_6\}$. (To recall: A *line* is a maximal totally ordered subset, a *cut* a maximal unordered one.) If however we insert one additional element into every cut, as illustrated in Fig. 10.5, the resulting structure then becomes K-dense.

One could be tempted to regard this as a meaningless trick, but this is not so. Recall that in Petri's double scale in Fig. 10.4, in many cases it *is* possible to read unambiguous values off both scales (only that it is *not always* so). These intermediate values were denoted by 5&6, etc. If we include these uncertain values into the consideration, and insert them so as not to violate the observable order relation, then we also get e.g. relationships of the form $5 \approx 5\&6 \approx 6$ and $4\&5 < 5\&6$. The resulting order structure corresponds exactly to the one in Fig. 10.5, if we name the elements appropriately.

This remarkable correspondence between measurement and process considerations can be carried even further: A record of the movement of the pointer over the double scale, from left to right, should be describable as a process in net terms.

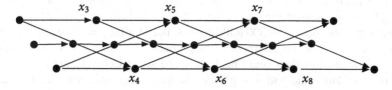

Fig. 10.5 Minimal K-dense reference scale. Arrows denote $<$, no connection denotes $\approx$

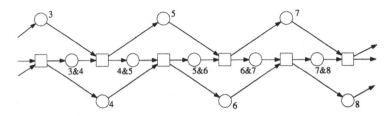

Fig. 10.6 Pointer moving along the double scale as a process

To this end, we show how to embed 10.5 in an occurrence net. At first sight it is clear that the order is not Dedekind complete. The gaps indicate where additional elements between the measurement values have to be inserted. These new elements then precisely represent the *events of passing* from one value reading to the next. As a result we obtain the occurrence net in Fig. 10.6, that is already known from other contexts, i.e., as a process of the Four Seasons system in Chap. 1.

10.9 Repercussions of Measurement Orders on Processes

In our discussion of the theory of measurement, the examples referred to the physical base units *mass* and *length*. The same questions however also appear in connection with the third fundamental base unit in physics—*time*, in particular in situations involving decisions, that rely on the temporal order of mutually independent events.

Example

In sports this is a typical question. The winner in a 100 meter race is determined by the order between different "concurrent" events of traversing the finish line (pun not intended). In this special case, the decision on temporal order can be reduced to a spatial comparison by means of a photograph of the finish.

Example

We encounter a significantly more complex situation in football, in particular with regard to the offside rule. In contrast to the 100 meter race, in the worst case here, two events have to be temporally ordered, which occur in spatially different locations: the passing of the ball, and the crossing of the offside line by another player. The problem has for instance been investigated by renowned eye specialists, and documented in the professional medical journal *Lancet*: The linesman must move his eye focus from one point to another. That takes time, typically about 250–300 milliseconds. In a skillfully arranged offside trap, this can lead to a misjudgment of magnitude 4.50 meters. For his decision, the linesman must then count back from the second observation to the point in time at which the ball was played. In sports such questions are of course especially critical, because of their far-reaching consequences. Victory or defeat may depend on knife-edge decisions.

In the length and weight measurements considered above, one of the main results was that we always have to take a ± 1-uncertainty into account. In a well-designed system, one will attempt to bound the magnitude of the deviation to a value that can be tolerated for the given context. It is interesting to note that the same approach can also be applied in temporal order comparisons between independent concurrent events.

Example

To illustrate, we consider a bus approaching a bus stop. If there is a passenger waiting, the driver has to stop and pick him/her up. If nobody is waiting, the driver may pass by without stopping. Now a client has complained to the bus company that he *was* waiting, but the bus nonetheless did not stop. The driver, on his side, contends that he had actually first noted the passenger in the rear-view mirror, because he, the passenger, definitely had arrived too late. Word stands against word, and the case will probably not be resolved.

But one thing *does* result: If there was a dispute about *this* bus, then the passenger is undoubtedly in time for the *next* one. Hence also in this case, there is a typical ± 1-problem lurking in the background. If the bus departures are sufficiently frequent, the negative consequences can be limited.

Example

To give another, more technical illustration, we recall the printer control in Fig. 1.9 in Chap. 1. Assume that it has been designed so that whenever agent a has issued a request (token on the condition a?), and the control unit is ready to accept it (token on c_a), then the printer will in fact be assigned to a. If however the control notes that the condition a? does not hold, the event l_a occurs and the control now offers the printer to agent b. The worst that can happen is that a is passed over *once*, when for example the request signal on a? has not yet stabilized.

If now the inner polling mechanism has been designed so that signals will stabilize within one cycle, we again get a ± 1-problem, which we can live with; probably also have to live with, since no decisively different solution seems viable: A system design without confusion is impossible, and the control must rely on a priority management that decides on the temporal order of the concurrent independent events involved in the confusion. (However we stress the point that these temporal relationships are only evaluated in the *immediate vicinity of a potential conflict.* This does not imply any assumption whatsoever of a global time concept.)

Petri himself did not explicitly elaborate on the consequences of his measurement approach for temporal order, one of the possible reasons being that he developed the theory at a time when he was still of the opinion that confusion in modeling could be avoided.

Fig. 10.7 Petri and global
time?

Let us return once more to the discussion between Dijkstra, Meisel and Petri on
the topic of the *glitch* in Sect. 6.5. All three of them were right: Such phenomena are
unavoidable by the laws of nature. It is only the consequences that can be minimized.
And as Petri remarked in that context, *confusion is the problem.*

The Prosperity Years

11

Contents

In the early 80s Petri was firmly seated in his saddle. His institute consisted of a sworn community of followers that certainly did not underestimate their importance. What happened there, more down to earth colleagues from other institutes often considered a glass bead game, divorced from reality (in the sense of Hermann Hesse). Petri's Institut für Systemforschung (ISF, Institute for System Research)—was often even stigmatized as the *I*nstitute for *S*cience *F*iction. Be that as it may, the institute was—to stay with the jargon—the *mothership of net theory*. There was an endless coming and going from guest researchers from all over the world, desiring to learn from the master. Because that was precisely what he had become: the undisputed authority and custodian of the insights into net theory. Many would call him the Pope of the nets, others however not, because, as was sometimes observed, the Pope still has someone above him.

His demeanor was renowned and feared. He could present his ideas for hours on end. If a co-worker was invited to a conversation in his office: "Could you spare me a minute?", you could cancel your intended day's schedule. If the invitation came before noon, you could be certain you would have to skip lunch. If it came in the afternoon, you had to postpone your evening arrangements.

You were sitting there at a small rectangular table, Petri held the cigarette in his left hand, the pencil in his right, explaining, with diagrams and minute writing, among other things net-theoretically based procedures for disarmament control, the mistakes in Bohr's understanding of quantum mechanics, the difference between information and entropy (basically only the opposite sign of the number values),

© Springer-Verlag Berlin Heidelberg 2015
E. Smith, *Carl Adam Petri*, DOI 10.1007/978-3-662-48093-9_11

Fig. 11.1 Petri in the mid 80s

or the reason to exclude side conditions in condition-event systems. The secretary Mrs. Kuhle repeatedly brought new supplies of coffee, until at a late hour she asked if she was still needed. Outside it became dark, but you did not have the courage to suggest to turn on the light. Every now and then Petri paused in the middle of a thought, remained silent and motionless for minutes, the cigarette glowing all by itself. Just before the ash column broke off, he continued his explanation at precisely the point he had stopped. It seems he used to develop many of his ideas in the speech pauses.

There was one certain co-worker however, whom Petri did not often invite to such conversations, because, as he said: "He always talks in monologue" and then adding with a thoughtful smile: "and, the fact is, that I myself like so much to play that part."

Ultimately however, these conversations were the best way to benefit from Petri's knowledge, and also often the only one; all in all he published his insights only sporadically. Also in conference lectures, you could never be sure which topic he would address. Together with the notes for a talk, he once packed away several overhead slides, with the remark: "In case there are questions." To the astonished, 'How could he know what questions would be asked', he had a simple reply: "I pretend to respond to the question, then show these slides and explain them."

Lecture Tours In the early 80s, Petri could not follow up the increasing number of all international invitations. Often he had himself apologized for health reasons, sometimes he declined with the explanation, the management board had not given him permission to travel. Prof. Norbert Szyperski, managing director of the GMD from 1981 to 1986, recalls: "When I met Mr. Petri, I occasionally might begin the conversation with the question, which trip I should deny him this time."

There was one country however, Petri took every opportunity to visit: China. He was fascinated by the people, the history, the culture, and the country's nature. Up to his old age he used to give lectures and courses there at various universities and

Fig. 11.2 Petri in China 1981

academies. In return, he invited Chinese guest researchers to the GMD time after time.

Paving the Way From the early 80s onward, various supporters played a substantial role in spreading the word of net theory world-wide, among them not least Grzegorz Rozenberg and Wolfgang Reisig.

Rozenberg, a Dutch computer scientist of Polish origin, was appointed professor at the University of Leiden in 1979, where he worked on automata theory and formal languages, amongst other things. He participated in the first Advanced Course 1979 in Hamburg, and quickly recognized the potential of the Petri nets. Because of his excellent scientific reputation, combined with his outstanding management skills, he soon advanced to become the undisputed leader of the net community. He organized conferences, was editor of publication series, for instance the yearly issue of *Advances in Petri Nets* published by the renowned Springer company; ultimately there was no major activity in the field of nets that did not benefit from his authoritative involvement.

During the annual international conference on Petri nets, a tutorial for beginners was also always arranged. Here Rozenberg had saved the "crown jewels" for himself, i.e., the introduction to the fundamental condition-event systems.

If you met him, say, at a conference, you could expect him to charge you with an *urgent* task, for instance to write a report on an article submitted to some journal: "The deadline is in two weeks." To the response, that presently you had *no time at all*, he showed himself sympathetic, pondered for a moment and decided: "Tell you what, let's make that *three* weeks."

Wolfgang Reisig was the author of the first standard text book on Petri nets. It appeared originally in German in the early 80s, and was also later translated into English, Italian, Polish, Chinese and Japanese. He joined the GMD in 1984, where he substantially influenced the development of the net approach towards practically oriented basic research, following the motto: "Nothing is as practical as a sound theory." With diplomatic dexterity he also promoted the spreading of nets

Fig. 11.3   *From left to right*: Wolfgang Reisig, Petri, Grzegorz Rozenberg

into professional circles who were traditionally more attached to other modeling techniques, by repeatedly giving lectures of the form: "What Petri nets and model X can learn from each other".

Even though he did not see his main focus in the discussion of Petri's physical-philosophical views, this author can confirm from his own experience that Reisig was very interested in it, and could contribute with many knowledgeable observations. The author is therefore very grateful that Wolfgang has agreed to accompany the present text with a foreword.

In 1988 Reisig accepted a professorship at the Technical University of Munich, where Wilfried Brauer had moved from the University of Hamburg some years earlier. In this way Munich became one the strongholds of net theory.

In 1993, after the fall of the Berlin wall and the reunification of Western and Eastern Germany, he moved to the Humboldt University in Berlin—in a sense also a kind of German-German-reunification, because that was where Peter Starke worked, the most prominent representative of nets in Eastern Germany.

Another important supporter of Petri's ideas at that time was Prof. Norbert Szyperski, member of the supervisory board of the GMD since 1979, managing director from 1981 to 1986, then chair of the supervisory board until the spring of 1991. Up to 1981 he held a chair in business administration at the University of Cologne. In 1994 he became founding president for the board of trustees of the Deutsche Forschungsnetz (DFN, German Research Network, a high-performance computer network).

Against this background, Szyperski in particular recognized the far reaching interdisciplinary relevance of Petri nets. In personal conversations with Petri and his co-workers from that time, this author got the impression that the years under Szyperski were amongst the most fruitful in the GMD.

In fact, down to the present day, Szyperski tries to promote Petri's ideas within his various activities in the advancement of innovative technologies.

11.1 Other Institute Activities

In 1982 the GMD was reorganized. Petri's Institute for System Research (ISF) became part of the newly founded "Institut für methodische Grundlagen" (IMG, Institute for Methodological Foundations), in which research fields with more long-term perspectives were subsumed. The IMG was given a collegial leadership, consisting of Petri, Prof. Fritz Krückeberg and Prof. Ulrich Trottenberg.

Krückeberg had been scientific-technological managing director and chairman of the executive board of the GMD until 1981, and now became executive leader in the triumvirate. He had been acquainted with Petri's works for a long time, and supported Petri wherever he could, in the institute management for instance by relieving him from administrative activities, for which Petri felt rather less inclined.

But he also supported the scientific development of Petri nets within his section of the institute; many of his co-workers developed methods for the algorithmic analysis of nets. In addition, the Krückeberg section worked on methods in interval arithmetic, a particular approach to error estimation in numerical computations.

The Trottenberg group consisted of mathematicians, who dealt with numerical algorithms, in particular with multigrid methods for the evaluation of differential equations.

From an organizational point of view the collaboration of the different institute sections worked smoothly, however the possibly hoped for scientific synergy remained limited. For details, the interested reader is referred to the already mentioned publication *Geschichte der GMD* [25].

Within his department of the institute, besides net theory, Petri also promoted various fields of research from which he expected important contributions to "communication with automata" in a broader sense. In the early days of the ISF, the focus was on software technology and data bases. Later the activities shifted towards the investigation of alternative computer architectures and computer-aided mathematics.

Alternative Computer Architectures Most of the computers in use today are built according to the von Neumann architecture, where program instructions are successively applied to data contained in memory cells. Basically, such an architecture is a realization of the operating principles of the previously proposed Turing machine.

Besides the Turing machine, there are also other models, which have the same computational power, but are based on entirely different principles. One of them is Alonzo Church's *lambda-calculus*, in which complex expressions for the computation of a function are successively reduced to simpler ones according to specific rules, until the result is obtained.

An approach to develop a corresponding *reduction machine*, directly based on this principle, was investigated in the Petri institute in the 70s and early 80s. Indeed, even a working prototype of such a machine was built, which did not go into production, however, because of the overwhelming competition of the von Neumann machines. Later the main protagonists, Klaus Berkling and Werner Kluge continued the theoretical work at the University of Syracuse in the United States and the University of Kiel, Germany.

Computer Algebra Until the mid 60s the use of computers in mathematics was limited to numerical computations. At that time the idea of computer algebra arose, in which the computer itself would perform symbolic formula conversions. In this way the computer enters the realm of real mathematics, and relieves the mathematician of necessary but tedious transformations. It becomes possible, for instance, to solve systems of partial differential equations symbolically with the aid of the machine. Already very early on Petri recognized the significance of this kind of communication with automata, and built up a research group at the institute. In particular, the group worked closely with the developer of the so-called Reduce system, Prof. Anthony Hearn from the University of Utah and collaborator at the Rand Corporation.

11.2   Time and Stochastic Processes

With the diffusion of nets, there came a desire to apply them in a great variety of domains. For instance, in the performance evaluation of system processes, there was an interest in using nets to determine numerical temporal relations. To this end, net classes were proposed, where time parameters were associated with the firing duration of transitions, or the stay of tokens on a place. The first in an ongoing series of international workshops on this topic of timed nets was arranged in Turin, Italy as early as 1985.

As sensible as such approaches may be in practice, for Petri they represented a downright contradiction to his fundamental view, that global time should not figure in the modeling of large systems. As in place-transition nets, precisely that which Petri attempted to remove from the theory, is instantly brought back in through the backdoor.

On occasion of the first conference on timed nets in Italy, he perhaps recalled the words of the great Italian poet Dante Alighieri: "Segui il tuo corso e lascia dir le genti"—follow your course and let the people talk.

Pursuant to his fundamental assumptions, he also had difficulties in sympathizing with stochastic approaches, where, for instance, conflicts are resolved probabilistically.

11.3   Advanced Course

As a follow-up to the first large advanced course on Petri nets in 1979 in Hamburg, a second one was arranged in 1986 in Bad Honnef, near Bonn. Petri gave two lectures: *Concurrency Theory* and *"Forgotten Topics" of Net Theory*. In the latter he addressed the topics most important to him personally, and distinguished himself from—in his opinion—hasty and thoughtless extensions of the net model. The course was well received, and occasionally the remark was heard: "There is nothing 'bad' about Honnef." (In reality "Bad" denotes that it is a spa town.)

The idea of the advanced courses was taken up in 1993 by a former co-worker of Petri's, César Fernández. He arranged a *Latin American Petri Nets Course '93* in Santiago de Chile. It was followed by the *Second International Course on Petri Nets for Latin America* in November 1995 in Campina Grande, Brazil.

11.4 Awards and Distinctions

When the Petri nets had definitively established themselves as an approved versatile modeling technique in the 1980s, Petri was honored with various prizes and decorations.

In 1988 he received the Officer's Cross of the Federal Republic of Germany.

In the same year, 1988, Petri was appointed honorary professor at the University of Hamburg. In a preliminary notification letter, he was told that, until the official appointment, he could use the title Professor "provisionally". Asked what that meant, Petri responded: "I cannot yet *demand* to be called 'Herr Professor', but no longer have to *object* if it happens." Before he *had* always objected, noting that he did not have any teaching obligations. Petri remained closely connected to the University of Hamburg until his passing away. In particular, he held advanced seminars on general net theory for many years.

In 1989 he was elected member of the Academia Europaea, shortly after it was founded. One central objective of the academy is to promote the public understanding of science. It has about 2000 members, among them 38 Nobel Prize winners.

Fig. 11.4 Petri with his wife Christel. Celebration in honor of his 60th birthday 1986 at the Birlinghoven Castle

In 1993 Petri received the Konrad Zuse Medal from the Gesellschaft für Informatik (GI, German Informatics Society). Since 1987 the prize has been awarded every 2 years to an outstanding German computer science researcher. It is the most renowned award in computer science in Germany. Petri's credentials were described as follows:

> Professor Dr. rer. nat. Carl Adam Petri is one of the leading, worldwide renowned and still active scientific pioneers of computer science. More than 30 years ago he began—far ahead of his time—to develop a general theory of discrete systems, based on the concepts of concurrency, distributedness, and asynchronous communication. Through Petri's imaginative contribution, a new area in computer science evolved, which incorporates parts of theoretical, practical and technical computer science, and also has many applications in other disciplines. An essential concept and modeling technique for distributed systems and parallel processes, which also denotes the whole of the research field, it bears the name of its inventor: the Petri net.

11.5   Werner von Siemens Ring

The decoration that emotionally moved him most, was undoubtedly the awarding of the Werner von Siemens Ring in 1997, a golden ring with emeralds and rubies. It is one of the highest awards for technical sciences in Germany, and is awarded about every 3 years. Some famous recipients are: Wernher von Braun, German and later US American rocket engineer; Walter Bruch, German television pioneer, who developed the PAL color-television system; Artur Fischer, the inventor of the Fischer S-plug; and the only other computer scientist, Konrad Zuse. Petri had been proposed by the ring recipient 1993, Prof. Eveline Gottzein, an engineer who like Petri originally came from Leipzig. She was an expert in load bearing and guide systems for high-speed magnetic railways. In her laudatory speech she said:

> With engineers, Petri nets led to a breakthrough in the treatment of discretely controlled systems. Petri nets play a pivotal role in the solution of design problems, because they for the first time permit an unambiguous description and also a convincing analysis of discretely controlled systems. Based on Petri nets, now it becomes possible to express system-invariants also for discrete control systems.

Present at the awarding ceremony was also the then German Federal President Roman Herzog, in his capacity as patron of the foundation council. On him, however, Petri's system approach did not seem to have made any lasting impression. As Petri recalls: "Formerly, Herzog had been the highest judge, caring for the Constitution. So I addressed him; I said that I had used the tool for which I was honored, for analyzing parliamentary and courtroom debate, and I hoped for a judgment. He did give it: 'Da hättense was Besseres tun können'—you might have used your time better. Said it, and turned away, and spoke to Mr. Bölkow, who had built planes."

11.6 Further Awards

In 1999 Petri received an honorary doctorate from the University of Zaragoza in Spain. The local press reported on the appointment under the heading "Karl Petri. El Filósofo de la Informática"; they had tacitly corrected the perceived misspelling in the name Carl.

In 2003 Petri was appointed Commandeur in de Orde van de Nederlandse Leeuw. The Order of the Dutch Lion was founded by King William I in 1815, and is the highest ranking civil order in the Netherlands. As grand master of the order, the king or queen carries out the appointment under the responsibility of the council of ministers.

In 2007 Petri was awarded an honors medal in gold for his life time achievement by the Academy of Transdisciplinary Learning and Advanced Studies (ATLAS).

In 2008 Petri received the renowned IEEE Computer Pioneer Award.

Decorations can however also move in the opposite direction. In honor of Petri, and named after him, the Carl Adam Petri Distinguished Technical Award is given once a year by the Society for Design and Process Science (SDPS). The first laureate in the year 2000 was Prof. Tadao Murata from the University of Chicago, who had substantially contributed to the dissemination of Petri nets in the United States.

The Late Years

12

Contents

Following a long and serious illness, Petri's wife Christel died in 1988. The two had been married for 32 years. According to his son Tobias, for about 5 years from 1987 onwards, Petri had to scale down his activities. He probably never really recovered from this loss.

In addition, at that time the GMD began to shift its policy away from basic research. In particular, Prof. Gerhard Goos, technical-scientific board member of the GMD from 1986 to 1991, was not really among Petri's admirers. Together with his mentor Friedrich Ludwig Bauer, Goos was the co-author of one of the first German computer science text books. In the GMD Goos had formerly been head of an institute for system technology, and of the Research Center for Program Structures in Karlsruhe. He seemed eager to mold the Petri institute after his ideas. In doing so, non-scientific motives may also on occasion have come to play a role. Anyway, the fact is that already for some time prior to Petri's retirement, the search for a successor from a *non-net-oriented* field was publicly advertised.

Petri himself commented on this peculiar form of research policy with a shrug: "After all, it is all about the election of a successor to the Chair Petri." Chinese guest researchers at the institute at that time noticed the rather subdued mood, and felt themselves reminded of their own so-called "hard years", when Deng Xiaoping led their country.

The successor elected to the Chair Petri then was Prof. Thomas Lengauer, who built up a productive research group on bioinformatics in the institute. However, Lengauer was not completely alien to nets: he had received a doctorate-degree for a thesis on Petri nets.

© Springer-Verlag Berlin Heidelberg 2015
E. Smith, *Carl Adam Petri*, DOI 10.1007/978-3-662-48093-9_12

12.1   Pensioner

At the age of 65 Petri retired from official professional life in the GMD in the year 1991.

However, he still considered his life's work to be far from complete. Amongst other things he intended to write a text book, in which he wanted to present his personal view on net theory. In fact, the present biography is based to a large extent on handwritten notes that Petri made during an extended vacation in Antalya, Turkey in 2002. As he observes there, his former superior in the GMD, Norbert Szyperski, had encouraged him to write such a monograph. Petri however rated his notes as much too rudimentary to be published at that time. Later, he chose to concentrate his declining strength on new tasks, not on looking back.

Private Scholar Formerly Petri used to write his manuscripts by hand, and to leave the typing to his secretary or occasionally also to his wife. Now, after 20 years of purely theoretical work, he decided once again to consider using computers to a practical effect, and set up a home office with PC. Installed on the computer there were the usual office applications, but also computer algebra programs, which he used extensively in his subsequent work.

The more he became acquainted with the possibilities of home computers, that had emerged in the meantime, the more he began to prepare his lectures and their presentations with the most modern tools, where his son Tobias often helped him with the graphics. Where he had initially planned to produce a book with an attached CD, Petri now instead began to directly produce CD-*lectures*, which he then also presented and distributed as such.

Teaching Activities Up until 2007 Petri continued his seminars in Hamburg. In addition, he frequently held lectures in various universities in China. He had had a

Fig. 12.1 Petri around 1991

special relationship with China for a long time already, and now he also found great pleasure in being able to spread his insights there. At the University of Beijing he held lectures before a large audience, which were recorded on video and thus also made available to the other Chinese universities.

Concerns about the medical impact of long journeys by air unfortunately prevented him from presenting his last series of CD-lectures himself, which he had prepared for the University of Xi'an at the beginning of 2008. He handed over the whole material to Prof. Rüdiger Valk from Hamburg, who had also been invited as visiting lecturer. Valk then presented the lectures for him. The CD from 2008 titled *On the Physical Basis of Information Flow* still exists today.

The Light Goes Out Petri's health was not subsequently to recover. Body and mind degraded continuously and irreversibly. Ten days before his 84th birthday, Carl Adam Petri passed away in Siegburg, close to his home. How was it that his son Tobias had described the painting by Konrad Zuse, hanging in Petri's office in the GMD, and later in his living room at home?: "A perennial with slender blossoms opening themselves to the light."

12.2 Net Theoretical Foundations of Physics

In his late works, Petri is especially concerned with the relationship between nets and physics, where he now mainly focuses on how *physical laws* can be represented in a discrete (Petri prefers the term *combinatorial*) form *in nets*. His special attention lies on the *Lorentz transformation*, which, in the framework of special relativity theory, describes the relationship between the determination of time and place of events, made by different observers. In addition, Petri wants to contribute to the resolution of the apparent *paradoxes in quantum mechanics* with his combinatorial approach.

He publishes his ideas for instance in the 1996 paper *Nets, Time and Space* in the journal *Theoretical Computer Science* [20]. A short version was issued in form of a CD-presentation *Petri's Belated Insights* in 2005. The same topic is also treated in the CD *Computing Nets Universe—dedicated to Konrad Zuse*, here with particular emphasis on his collaboration with Zuse. In 2007 he writes an article *Das Universum als großes Netz* (The Universe as a Large Net) in a special edition, dedicated to Zuse, of the journal *Spektrum der Wissenschaft* (the German edition of the renowned popular science magazine *Scientific American*) with the general theme: *Is the universe a computer?*

Since his early encounter with the works of Einstein in the National Library in Leipzig, Petri was fascinated by Einstein's radical questioning of apparently obvious truths. For Petri, Einstein's physics always remained the reference framework.

Petri later even spoke of "physics as the birthplace of net theory". Throughout his scientific life, he felt sorry that even though net theory had found many different successful applications, it had however not done so at its "place of birth". As mentioned, he now took the initiative himself. He did not want to go as far as

Fig. 12.2 Petri around 2000

Zuse, who saw physics as an incarnation of information processing. What he wanted however, was to trace back the laws of physics to their combinatorial discrete nature, as he had already done in the case of density and continuity, as well as in the field of measurement.

Presumably he also felt a certain gratitude: physics had given him so much, now it was his turn to fertilize physics. This was so important to him that the other leitmotiv of his work, the development of a formal pragmatics, did not perhaps receive its due share of attention.

12.3   Formal Pragmatics

The effort to develop a formal pragmatics can be traced throughout Petri's work. Some introductory ideas can be found in the article *The Pragmatic Dimension of Net Theory* [21], but unfortunately he did not manage to develop a comprehensive theory during his lifetime.

A central issue in formal pragmatics is to study the purpose and effect of communication. For instance, in his theory of measurement, Petri argues that measurement values only make sense in the context *for which* they were established. A measurement's information content is determined by the number of possible alternative actions, from which one is then chosen, depending on the outcome of the measurement.

In *Communication Disciplines* [17], Petri emphasizes that copying does not modify a message itself, but very possibly its significance for the persons involved. A secret that is copied is not a secret anymore. In fact, in the communication disciplines, pragmatic concepts are visible throughout, for instance *authority, change of mind, formation of thought models, power to direct.*

The term "communication with automata" in Petri's thesis thus actually obtains still another meaning: The use of computers is controlled by rule-based communication, which however also includes the rules of purposeful communication *between* human agents.

Code of Civil Procedure As a prototypical example of rule-based communication between individuals and groups of individuals, in the early 80s Petri initiated a research project to model and analyze the German Code of Civil Procedure in the framework of net theory. The project was among others based on preliminary work by Petri's long time colleague Anatol Holt, who had published a much-noticed article *Petri Nets and Legal Systems* in an American professional juristic journal in 1971.

In the analysis of the civil procedures, it turned out that the formal treatment of e.g. interests, responsibility, deadlines, is perfectly possible with nets, where, however, special attention has to be given to the formalization of the modal concepts *must, should, may not.*

Subsequently, these endeavors were unfortunately not pursued much further. One of the reasons was perhaps that such apparent esoteric topics did not attract wholehearted support from the current leadership in the GMD. According to Petri, another reason was that he, as was very common, had kept all the preparatory work only in his head until it was ready to be published, where, as he said, it unfortunately was lost following an illness, an inflammation of the middle ear.

Logics Petri nonetheless further investigated selected topics in pragmatics. Most prominent of these were different kinds of formal logics and their relationships to net theory. *Modal logic*, for instance, deals with the concepts *possible* and *necessary.* Here it could be demonstrated that the case classes in condition-event systems can be interpreted as natural models of modal logic, in the sense of the widely acknowledged *Kripke semantics.*

The same essentially also holds true for the *deontic interpretation* of the modal operators in the sense of *may* and *must.* In the investigations of the Code of Civil Procedure it became clear, however, that *must* is not immediately representable in nets, because it is always accompanied by a temporal *deadline*, within which the obligation has to be met.

As a particularly promising contribution to formal pragmatics, Petri saw the so-called *game semantics* (as it is nowadays usually called in English), initially developed by the German philosopher and mathematician Paul Lorenzen under the name "dialogische Logik" (dialogical logic) in the late 1950s. It is closely based on the model of human reasoning.

The underlying idea can be illustrated by a traveler R arriving at customs control with two suitcases 1 and 2. The customs officer asks him to open suitcase 1. R opens it, but it does not contain any smuggled goods, and R may pass. A control is after all a random sampling. The cigarettes were in suitcase 2.

If the customs officer had asked: Please open *one* of the suitcases, and *R* had opened suitcase 1, the immediate result would have been the same, but the pragmatic status would have been different; it would no longer have been a random sampling.

By the same token, Petri traced back the difference between the logical operators AND and OR to a difference in pragmatic context: A criminal has chosen one of the escape routes *A or B*. The police however do not know which one, and therefore must follow *both A and B*.

It is only to be deplored that Petri did not develop his formal pragmatics into a comprehensive theory. He himself had perhaps commented upon it like this: We take the pole star as a direction indicator, even if we do not reach it.

12.4   Closing Remarks

Petri nets owe their worldwide success to their adaptability to a huge range of applications. As we have seen, Petri himself is primarily interested in the understanding of fundamental phenomena and relationships, and only then, based on this solid knowledge, in the modeling of concrete systems. Comparing the widespread applications with Petri's personal work, it is in fact not always easy to recognize that they all stem from a common basic idea.

On closer inspection, though, the common root of all variants of net theory and its applications becomes clear, namely the *asynchronous, discrete* modeling that, arguably, as thoroughly as no other approach has made it possible to draw a sharp distinction between causal *dependence* and *independence*. Only in this way could an instrument arise, which keeps its due distance from classical ideas based on totally ordered structures, the metrics of time and on the real-valued continuum.

Indeed, all modified approaches to nets, such as *timed* or *stochastic* nets, are ultimately also solidly founded on the underlying causality structure inherent in nets. This does, in fact, permit a two-stage analysis method: The causal level always remains visible, simply because the objects are Petri nets. Determinations of time or probabilities then take place in a second level above it.

Automata Theory and Physics It is not uncommon that a theoretical model proposed in one context, then actually gains much greater popularity elsewhere. Warren McCulloch and Walter Pitts attempted to investigate the capability of the human brain with their neuronal model; however, their ideas mainly caught on within automata theory in computer science. Noam Chomsky wanted to understand the deep structure of human languages, but his approaches were more successful in the syntactic description of *programming languages*.

With Petri the situation is similar, or, in truth, rather the opposite. In his thesis he wanted to make a contribution to automata theory. The approach initially found acceptance in quite different contexts, however. A large part of his later work was devoted to the physical foundations of information processing, but the applications freed themselves, and, due to the expressive power of the model, moved on to other objectives.

As already mentioned, Petri felt sorry in his final years that his nets had not yet been accepted and put into use in physics, or, as he termed it at their "place of birth". He decided to take the matter into his own hands. The physicists could not, unfortunately, be convinced. They probably had the same attitude that a court of law has, when a layman bases his argument directly on the constitution. In his conversations with physicists, such as Richard Feynman, John Stewart Bell, Gerard 't Hooft, the agreement did not usually reach beyond one of mutual recognition and respect.

Carl Adam Petri, the Person In general, Petri was little concerned with the day-to-day life of professional scientific circles. In his youth he was often timid and fearful, but then grew almost too rapidly into the role of a master teacher beyond the trifles of the world, with an entourage of devoted acolytes, who never doubted his wisdom. To the outer world he gave much advice, but he never fought to defend his opinions according to the conventions of the scientific community.

Taken to an extreme, one could say that the nets did not succeed *because* of Petri, but in spite of him. But this misses the point. It was precisely for the reason that he was an autodidact in many fields, that he did not develop any undue respect towards scientific authorities. Presumably this is the only way to break with outdated concepts. Most of the time such an attitude goes awry, but in Petri's case it did lead to success.

In addition to the particular theories he addressed, what was especially remarkable about Petri, was that he was willing to question even apparently unimportant minor details, and test them over and over again. That was always what was fascinating in his works and lectures: You read, you listened, you did not understand everything, but then suddenly the realization dawned upon you: Yes, of course, that is the way it is. How come I did not see it like this before?

Fig. 12.3 Carl Adam Petri

Fig. 12.4 Outlook

Outlook Many of the questions raised by Petri will probably have to wait a long time for satisfactory answers, e.g. the development of a finitary mathematics that directly reflects the objectives of the users; or measurement methods that critically scrutinize historically grown conventions, as a necessary precondition for an adequate theory based on practice; or the overall context of information processing, including the interests and obligations of the agents involved; or the unified approach to communication between different entities, be they humans, groups of individuals or machines.

Petri wanted to radically reform computer science, or more precisely, he rather considered it still to be in its early infancy, and wanted to make a contribution that would guide it in the right direction in its development phase and youth. In one of his seminal articles *Non-Sequential Processes* he notes: "The history of geometry is measured in thousands of years, that of computer science in decades. Perhaps coming generations will compare the computer scientists of today with the surveyors of ancient Egypt who went about their work in the mud of the Nile, equipped with rules and tools just sufficient for the purpose at hand."

Bibliography

1. E. Best, C. Fernández, *Nonsequential Processes. A Petri Net View* (Springer, New York, 1989)
2. C. Fernández, *Net Topology I, II.* St. Augustin: Gesellschaft für Mathematik und Datenverarbeitung, Internal Report ISF-75–09, 76–02 (1975/6)
3. H.J. Genrich, Ein Kalkül des Planens und Handelns. *Berichte der GMD 111: Ansätze zur Organisationstheorie rechnergestützter Informationssysteme* (R. Oldenbourg Verlag, München, Wien, 1979), pp. 77–92
4. C.A. Petri, *Kommunikation mit Automaten* (Institut für Instrumentelle Mathematik, Schriften des, Bonn, 1962), IIM Nr. 2
5. C.A. Petri, *Communication with Automata.* Griffiss Air Force Base, New York, Technical Report RADC-TR-65-377, Vol 1, Suppl. 1, translation of [4] (1966)
6. C.A. Petri, Fundamentals of a Theory of Asynchronous Information Flow. *Proc. of IFIP Congress 62* (North Holland, Amsterdam, 1963), pp. 386–390
7. C.A. Petri, Grundsätzliches zur Beschreibung diskreter Prozesse. *3. Colloquium über Automatentheorie, Hannover 1965* (Birkhäuser-Verlag, Basel, 1967), pp. 121–140
8. C.A. Petri, *Fundamentals of the Representation of Discrete Processes* Gesellschaft für Mathematik und Datenverarbeitung, St. Augustin, ISF-Report 82.04, translation of [7] (1982)
9. C.A. Petri, Concepts of Net Theory. *Mathematical Foundations of Computer Science: Proc. of Symposium and Summer School, High Tatras, Sep. 3–8, 1973* (Math. Inst. of the Slovak Acad. of Sciences, Bratislava, 1973), pp. 137–146
10. C.A. Petri, *Interpretations of Net Theory* Gesellschaft für Mathematik und Datenverarbeitung, St. Augustin, Internal Report ISF-75–07 (1976)
11. C.A. Petri, *Nicht-sequentielle Prozesse* (Gesellschaft für Mathematik und Datenverarbeitung, St. Augustin, Internal Report ISF-76-6 (1976)
12. C.A. Petri, *Non-Sequential Processes* (Gesellschaft für Mathematik und Datenverarbeitung, St. Augustin, Internal Report ISF-77-01, translation of [11] (1977)
13. C.A. Petri, General Net Theory. *Computing System Design: Proc. of the Joint IBM University of Newcastle upon Tyne Seminar, Sep. 1976,* pp. 131–169 (1977)
14. C.A. Petri, Communication Disciplines. *Computing System Design: Proc. of the Joint IBM University of Newcastle upon Tyne Seminar, Sep. 1976* , pp. 171–183 (1977)
15. C.A. Petri, Modeling as a Communication Discipline. Unpublished manuscript, abstract in: *Measuring, Modeling and Evaluating Computer Systems* (North Holland, Amsterdam, 1977), p. 435
16. C.A. Petri, *Concurrency as a Basis of System Thinking* Gesellschaft für Mathematik und Datenverarbeitung, St. Augustin, Internal Report ISF-78-06 (1978)
17. C.A. Petri, Kommunikationsdisziplinen. *Berichte der GMD 111: Ansätze zur Organisationstheorie rechnergestützter Informationssysteme* (R. Oldenbourg Verlag, München, Wien, 1979), pp. 63–76
18. C.A. Petri, State-transition structures in physics and in computation. Int. J. Theor. Phys. **21**(12), 979–992 (1982)
19. C.A. Petri, E. Smith, Concurrency and Continuity. *Lecture Notes in Computer Science Vol. 266: Advances in Petri Nets 1987* (Springer, New York, 1987), pp. 273–292

© Springer-Verlag Berlin Heidelberg 2015

E. Smith, *Carl Adam Petri*, DOI 10.1007/978-3-662-48093-9

20. C.A. Petri, Nets, time and space. Theor. Comput. Sci. **153**, 3–48 (1996)
21. C.A. Petri, E. Smith, The pragmatic dimension of net theory. J. Integ. Des. Process Sci. Trans. SDPS **2**(1), 1–7 (1998)
22. W. Reisig, *Petrinetze. Modellierungstechnik, Analysemethoden, Fallstudien* (Vieweg-Teubner Verlag, Berlin, 2010)
23. W. Reisig, *Understanding Petri Nets. Modeling Techniques, Analysis Methods, Case Studies* (Springer, New York, 2013)
24. E. Smith, On the border of causality: contact and confusion. Theor. Comput. Sci. **153**, 245–270 (1996)
25. J. Wiegand, *Informatik und Großforschung. Geschichte der Gesellschaft für Mathematik und Datenverarbeitung* (Campus Verlag, Frankfurt, 1994)

Index

© Springer-Verlag Berlin Heidelberg 2015
E. Smith, *Carl Adam Petri*, DOI 10.1007/978-3-662-48093-9

Printed in the United States
By Bookmasters

Printed in the United States
By Bookmasters